AF598139

METHODS IN MOLECULAR BIOLOGY™

Series Editor
John M. Walker
School of Life Sciences
University of Hertfordshire
Hatfield, Hertfordshire, AL10 9AB, UK

For further volumes:
http://www.springer.com/series/7651

Bioluminescent Imaging

Methods and Protocols

Edited by

Christian E. Badr

Experimental Therapeutics and Molecular Imaging Laboratory, Department of Neurology, Neuroscience Center, Massachusetts General Hospital, Boston, MA, USA; Program in Neuroscience, Harvard Medical School, Boston, MA, USA

Editor
Christian E. Badr
Experimental Therapeutics and Molecular Imaging Laboratory,
Department of Neurology, Neuroscience Center,
Massachusetts General Hospital, Boston, MA, USA
Program in Neuroscience, Harvard Medical School, Boston, MA, USA

ISSN 1064-3745 ISSN 1940-6029 (electronic)
ISBN 978-1-62703-717-4 ISBN 978-1-62703-718-1 (eBook)
DOI 10.1007/978-1-62703-718-1
Springer New York Heidelberg Dordrecht London

Library of Congress Control Number: 2013950898

Printed on acid-free paper

Humana Press is a brand of Springer
Springer is part of Springer Science+Business Media (www.springer.com)

Preface

In the deep, dark ocean, many living organisms orchestrate one of the most amazing light shows the eye can ever witness. Obscured by total darkness and surrounded by predators, these organisms have evolved and acquired light-emission properties, used for self-defense and communication with the environment. These "glow-in-the-dark" species can also be found on land and are limited to fungi, bacteria, and insects such as fireflies.

Since the early years, light-emitting species have been a source of fascination and have stemmed the curiosity of great thinkers, philosophers, as well as scientists. Aristotle (384–322 bc) wrote one of the first detailed observations on auto-luminescent organisms and used the term "cold fire" in reference to light produced by these species, which, unlike fire, is not associated with a temperature increase. In his *Naturalis Historia*, Pliny the Elder (23–79 ad) gave a detailed description of several bioluminescent species including fireflies, glowworms, and jellyfish. Fast-forwarding to the late nineteenth century, the French physiologist Raphael Dubois made early discoveries on firefly bioluminescence. Working with grinded firefly abdomens, he described two essential components of the light emission properties, luciferase and luciferin, which he named after the fallen angel of light, Lucifer (from the Latin lux: light, and ferre: to bring). In the last half-century, the scientific interest in bioluminescent species grew beyond "the how and the why" and much attention shifted towards reproducing these luminescent properties in a test tube. This started a new field of bioluminescence imaging with endless applications that are now used to illuminate the darker side of scientific research.

This book represents a distillation of protocols and methodologies that use bioluminescence imaging as a tool for visualizing and tracking various biological processes. It covers a wide spectrum of methods and tools thoroughly described and illustrated through various biological applications covering diverse fields such as cellular and molecular biology, oncology, neurology, infectious diseases, immunology, and others. The first chapter is a brief introduction meant to define and describe different bioluminescence reporters and their properties. A significant part of this chapter is dedicated to the limitations and drawbacks of bioluminescence imaging. This part might prove useful for designing and choosing the appropriate reporter for your assay and defining what is attainable or not with bioluminescence imaging. Succeeding chapters are arranged by topic and describe practical procedures and applications of different bioluminescent reporters, from photoproteins (Aequorin) to bacterial luciferases as well as other secreted (such as Gaussia) and nonsecreted luciferases (such as Firefly).

I owe a profound thank you to all of the authors who contributed to this book. It was a real pleasure to work with them. I am grateful to Professor John Walker for his guidance and valuable input. I am deeply thankful for Dr. Bakhos Tannous for his thoughtful advice and critical assessment of this book. Finally, my most special thank you goes to Carla, my wife, for all her patience and infinite support.

It is my hope that this book provides diverse and comprehensive techniques to everyone interested in implementing bioluminescence-based imaging in their laboratory, regardless of their previous level of experience with such methodologies.

Boston, MA, USA ***Christian E. Badr***

Contents

Contributors

ROMAIN J. AMANTE • *Experimental Therapeutics and Molecular Imaging Laboratory, Neuroscience Center, Department of Neurology, Massachusetts General Hospital, Boston, MA, USA; Program in Neuroscience, Harvard Medical School, Boston, MA, USA*

JAMIL AZZI • *Renal Division, Transplantation Research Center, Brigham and Women's Hospital and Children's Hospital, Harvard Medical School, Boston, MA, USA*

CHRISTIAN E. BADR • *Experimental Therapeutics and Molecular Imaging Laboratory, Department of Neurology, Neuroscience Center, Massachusetts General Hospital, Boston, MA, USA; Program in Neuroscience, Harvard Medical School, Boston, MA, USA*

SAMANTHA BLAZQUEZ • *Plate-Forme d'Imagerie Dynamique, Imagopole, Institut Pasteur, Paris, France*

M. SARAH S. BOVENBERG • *Experimental Therapeutics and Molecular Imaging Laboratory, Department of Neurology, Neuroscience Center, Massachusetts General Hospital, Boston, MA, USA; Program in Neuroscience, Harvard Medical School, Boston, MA, USA; Department of Neurosurgery, Leiden University Medical Center, Leiden, The Netherlands*

XANDRA O. BREAKEFIELD • *Department of Neurology, Neuroscience Center, Massachusetts General Hospital, Boston, MA, USA; Department of Radiology, Center for Molecular Imaging Research, Massachusetts General Hospital, Boston, MA, USA; Program in Neuroscience, Harvard Medical School, Boston, MA, USA*

MIGUEL A.S. CAVADAS • *Systems Biology Ireland, University College Dublin, Dublin, Ireland*

ALEX CHEONG • *Systems Biology Ireland, University College Dublin, Dublin, Ireland*

JUNJI CHIDA • *Division of Enzyme Chemistry, Institute for Enzyme Research, The University of Tokushima, Tokushima, Japan*

EUIHEON CHUNG • *Department of Medical System Engineering and School of Mechatronics, Gwangju Institute of Science and Technology, Gwangju, South Korea*

DAN CLOSE • *Biosciences Division, Oak Ridge National Laboratory, Oak Ridge, TN, USA; The Joint Institute for Biological Sciences, The University of Tennessee, Knoxville, TN, USA*

M. HANNAH DEGELING • *Experimental Therapeutics and Molecular Imaging Laboratory, Department of Neurology, Neuroscience Center, Massachusetts General Hospital, Boston, MA, USA; Program in Neuroscience, Harvard Medical School, Boston, MA, USA; Department of Neurosurgery, Leiden University Medical Center, Leiden, The Netherlands*

JOE DRAGAVON • *Plate-Forme d'Imagerie Dynamique, Imagopole, Institut Pasteur, Paris, France*

SCOTT J. GOLDMAN • *U.S. Army Research Institute of Environmental Medicine, Natick, MA, USA*

UWE HIMMELREICH • *Department of Imaging and Pathology, Biomedical MRI/MoSAIC, KU Leuven, Leuven, Belgium*

SATOSHI INOUYE • *Department of Biochemistry, School of Dentistry, Aichi-Gakuin University, Nagoya, Japan; Yokohama Research Center, JNC Corporation, Yokohama, Japan*

SHENGKAN JIN • *Department of Pharmacology, Rutgers University-Robert Wood Johnson Medical School, Piscataway, NJ, USA*

HIROSHI KIDO • *Division of Enzyme Chemistry, Institute for Enzyme Research, The University of Tokushima, Tokushima, Japan*
SOONHAG KIM • *Laboratory of Molecular Imaging, Department of Biomedical Science, College of Life Science, CHA University, Seoul, Republic of Korea*
SUNG BAE KIM • *Research Institute for Environmental Management Technology, National Institute of Advanced Industrial Science and Technology (AIST), Tsukuba, Japan*
HAE YOUNG KO • *Laboratory of Molecular Imaging, Department of Biomedical Science, College of Life Science, CHA University, Seoul, Republic of Korea*
SOŇA KUCHARÍKOVÁ • *Laboratory of Molecular Cell Biology, Department of Molecular Microbiology, Institute of Botany and Microbiology, VIB, KU Leuven, Leuven, Belgium*
CHARLES P. LAI • *Department of Neurology, Neuroscience Center, Massachusetts General Hospital, Boston, MA, USA; Program in Neuroscience, Harvard Medical School, Boston, MA, USA*
BERTRAND LAMBOLEZ • *Neurobiologie des processus adaptatifs, UMR7102, Université Pierre et Marie Curie, Paris, France*
MATTHEW B. LAWRENZ • *Department of Microbiology and Immunology, Center for Predictive Medicine for Biodefense and Emerging Infectious Diseases, University of Louisville School of Medicine, Louisville, KY, USA*
YOUNG SIK LEE • *College of Life Sciences and Biotechnology, Korea University, Seoul, Republic of Korea*
GRANT K. LEWANDROWSI • *Experimental Therapeutics and Molecular Imaging Laboratory, Department of Neurology, Neuroscience Center, Massachusetts General Hospital, Boston, MA, USA; Program in Neuroscience, Harvard Medical School, Boston, MA, USA*
GARY D. LUKER • *Center for Molecular Imaging, Department of Radiology, University of Michigan, Ann Arbor, MI, USA; Departments of Biomedical Engineering, University of Michigan, Ann Arbor, MI, USA; Department of Microbiology and Immunology, University of Michigan, Ann Arbor, MI, USA*
KATHRYN E. LUKER • *Center for Molecular Imaging, Department of Radiology, University of Michigan, Ann Arbor, MI, USA; Departments of Biomedical Engineering, University of Michigan, Ann Arbor, MI, USA; Department of Microbiology and Immunology, University of Michigan, Ann Arbor, MI, USA*
CIARA N. MAGEE • *Renal Division, Transplantation Research Center, Brigham and Women's Hospital and Children's Hospital, Harvard Medical School, Boston, MA, USA*
CASEY A. MAGUIRE • *Department of Neurology, Massachusetts General Hospital, Boston, MA, USA; Neuroscience Program, Harvard Medical School, Boston, MA, USA*
MARWAN MOUNAYAR • *Renal Division, Transplantation Research Center, Brigham and Women's Hospital and Children's Hospital, Harvard Medical School, Boston, MA, USA*
DAN T. NGUYEN • *Department of Medical System Engineering and School of Mechatronics, Gwangju Institute of Science and Technology, Gwangju, South Korea*
JONAS NILSSON • *Department of Radiation Sciences, Oncology, Umeå University, Umeå, Sweden*
ABDESSALEM REKIKI • *Plate-Forme d'Imagerie Dynamique, Imagopole, Institut Pasteur, Paris, France*
STEVEN RIPP • *The Center for Environmental Biotechnology, The University of Tennessee, Knoxville, TN, USA*
KELLY L. ROGERS • *The Walter and Eliza Hall Institute of Medical Research, Parkville, VIC, Australia*
CHELSEA SAMSON • *Vanderbilt School of Medicine, Nashville, TN, USA*

GARY SAYLER • *The Joint Institute for Biological Sciences and The Center for Environmental Biotechnology, The University of Tennessee, Knoxville, TN, USA*
SPENCER SHORTE • *Plate-Forme d'Imagerie Dynamique, Imagopole, Institut Pasteur, Paris, France*
TAKAHIRO SUZUKI • *Department of Biochemistry, School of Dentistry, Aichi-Gakuin University, Nagoya, Japan; Yokohama Research Center, JNC Corporation, Yokohama, Japan*
BAKHOS A. TANNOUS • *Experimental Therapeutics and Molecular Imaging Laboratory, Department of Neurology, Neuroscience Center, Massachusetts General Hospital, Boston, MA, USA; Program in Neuroscience, Harvard Medical School, Boston, MA, USA*
MARIE TANNOUS • *Notre Dame University, Barsa, Lebanon*
IOANNA THEODOROU • *Plate-Forme d'Imagerie Dynamique, Imagopole, Institut Pasteur, Paris, France*
RÉGIS TOURNEBIZE • *Plate-Forme d'Imagerie Dynamique, Imagopole, Unité INSERM U786, Institut Pasteur, Paris, France*
LUDOVIC TRICOIRE • *Neurobiologie des processus adaptatifs, UMR7102, Université Pierre et Marie Curie, Paris, France*
PATRICK VAN DIJCK • *Laboratory of Molecular Cell Biology, Department of Molecular Microbiology, Institute of Botany and Microbiology, VIB, KU Leuven, Leuven, Belgium*
SJOERD VAN RIJN • *Department of Neurosurgery, Cancer Center Amsterdam, Neuro-oncology Research Group, VU University Medical Center, Amsterdam, The Netherlands*
GREETJE VANDE VELDE • *Department of Imaging and Pathology, Biomedical MRI/ MoSAIC, KU Leuven, Leuven, Belgium*
JONATHAN M. WARAWA • *Department of Microbiology and Immunology, Center for Predictive Medicine for Biodefense and Emerging Infectious Diseases, University of Louisville School of Medicine, Louisville, KY, USA*
THOMAS WÜRDINGER • *Department of Neurosurgery, Cancer Center Amsterdam, VU University Medical Center, Neuro-oncology Research Group, Amsterdam, The Netherlands; Department of Neurology, Neuroscience Center, Massachusetts General Hospital and Neuroscience Program, Harvard Medical School, Boston, MA, USA*
TINGTING XU • *The Joint Institute for Biological Sciences and The Center for Environmental Biotechnology, The University of Tennessee, Knoxville, TN, USA*
HIROSHI YAMASHITA • *Department of Medical System Engineering and School of Mechatronics, Gwangju Institute of Science and Technology, Gwangju, South Korea*

Chapter 1

Bioluminescence Imaging: Basics and Practical Limitations

Christian E. Badr

Abstract

Over the last three decades, imaging has been a thriving field with continuous egression of more reliable and highly sophisticated tools and techniques allowing better understanding of biological processes in living organisms. This field continues to expand and its applications broaden to encompass limitless applications in various biomedical research areas. It is however, of utmost importance to understand the capabilities and limitations of this technique as new challenges and hurdles continue to arise. This chapter describes the general properties of bioluminescence imaging and commonly used reporters while underlining the challenges and limitations with these modalities.

Key words Bioluminescence, Imaging, Photoprotein, Lux, Luciferase

1 Introduction

Imaging technologies emerged at the beginning of the twentieth century as a way to complement morphological observations. Molecular imaging (MI) allows a visual (often quantitative) study of molecular, cellular, biochemical, and physiological processes in respect to space and time in a living organism. Advances in molecular and cellular biology, discovery and design of new reporter proteins and molecular probes, as well as the use of transgenic animals have contributed greatly to the expansion of the imaging field.

Imaging systems can be classified under three groups [1, 2]: the energy used to obtain the visual information (X-rays, positrons, photons, or sounds waves); the spatial resolution (macroscopic, mesoscopic, or microscopic); or the type of information acquired (anatomical, physiological, cellular, or molecular). Macroscopic imaging techniques such as computed tomography (CT), magnetic resonance imaging (MRI), and ultrasound are well established in the clinic and provide anatomical and physiological information. Despite the progress in the imaging field, current techniques have not yet been clinically optimized to provide

Christian E. Badr (ed.), *Bioluminescent Imaging: Methods and Protocols*, Methods in Molecular Biology, vol. 1098, DOI 10.1007/978-1-62703-718-1_1, © Springer Science+Business Media New York 2014

detailed information on specific molecular events (changes in gene expression, activation of certain signaling networks, etc.), especially in the context of the disease (e.g., comparing normal cells versus tumor cells). Hence, new techniques that provide molecular information are currently under development and some are starting to emerge in both preclinical and clinical settings. The most commonly used molecular imaging modalities include positron emission tomography (PET), single-photon-emission CT (SPECT), magnetic resonance (MR), fluorescence-mediated tomography (FMT), laser-scanning confocal microscopy, and bioluminescence imaging (BLI).

Molecular imaging allows a spatiotemporal determination of the molecules of interest as well as monitoring of a specific biological process in living cells or in animals. The applications of this field are numerous and include:

- Localization and trafficking of proteins through different cellular compartments using fluorescent reporters fused to the protein of interest.
- Monitoring of gene expression using DNA-binding responsive elements acting as promoters to study the expression of certain genes [3].
- Visualization of enzymatic activity such as proteases by inserting a small peptide in the middle of the reporter which is recognized and cleaved by the enzyme [4] or LC3 cleavage during autophagy [5, 6].
- Monitoring of cell trafficking or cell migration such as stem cells, T-lymphocytes and tracking of cancer cell metastasis, migration, or invasion [7–11].

Molecular imaging, in particular optical imaging (fluorescence or bioluminescence), is also commonly used in drug discovery. Once identified, a molecule hit is validated in animal models using noninvasive imaging techniques to study the drug biodistribution, pharmacokinetics, and potency [12]. In cancer, tumor growth and response to various compounds or gene therapy is commonly monitored using bioluminescence imaging [13, 14]. Unlike histopathological and cytopathological studies, molecular imaging techniques require no chemical fixation or isolation of tissues and organs rendering the study of physiological processes over time in the same biological sample possible. These processes can be determined in their own biological context, while abrogating the need to sacrifice the experimental animals. Data obtained could often be quantified using designed software, which translates the signal into numerical measures.

Bioluminescence imaging offers powerful and versatile tools for monitoring of different biological processes in cultured cells and in living animals. This technique had become indispensable in many molecular biology laboratories, with a diverse and broad range of applications encompassing various biomedical fields and preclinical research areas.

2 Bioluminescent Reporters

Bioluminescence (BL) is the natural production of light often seen in different lower organisms (beetles, bacteria, algae, crustaceans, annelids, mollusks, and coelenterates). Numerous bioluminescent systems exist in nature, many of which have been isolated and studied in laboratories and the biochemical properties of their light emission properly defined. Luminescence is generated through a chemical reaction where the enzyme (luciferase) oxidizes a substrate (luciferin) leading to photon emission. Some luciferases require the presence of cofactors (ATP, Mg^{2+}) for their activity.

Fluorescence is another widely used optical imaging modality that also generates light through a chemical reaction. Unlike BL, this light generation is triggered by an external light source. These two imaging modalities also differ by the signal intensity as well as the signal-to-noise (S/N) ratio. Although fluorescent signals are usually brighter than bioluminescence, the background noise due to autofluorescence is also higher. The high sensitivity of bioluminescence is mostly due to a virtually absent background yielding higher S/N ratios.

Bioluminescent reporters can be divided into two major groups: photoproteins and luciferases. Photoproteins emit light in proportion to the concentration of the protein itself, while in a luciferin-luciferase reaction, photon emission is directly proportional to the amount of luciferin [15].

2.1 Photoproteins

This family encompasses proteins that emit light in proportion to the protein itself and do not require an enzyme (luciferase) [15]. While coelenterate photoproteins have been notorious and widely employed as sensitive reporters for Ca^{2+} detection, not all photoproteins are Ca^{2+} sensitive. In fact, other photoproteins, which activity depends on the presence of H_2O_2, ATP, Mg^{2+} or superoxide, have been reported [16]. Most studies employ the Ca^{2+}-regulated photoproteins that use coelenterazine as a substrate (aequorin, obelin, phialidin, berovin). Aequorin, from the jellyfish *Aequorea victoria* is the best-known and widely used photoprotein. It is commonly used to monitor Ca^{2+} concentrations from a single cell [17]. Like all coelenterate photoproteins, aequorin has an approximate molecular mass of 20 kDa and emits blue light in the presence of Ca^{2+} [15]. The aequorin protein contains coelenterazine in the central cavity and is capable of binding Ca^{2+} through its three "EF hand" motifs [18]. Upon binding to Ca^{2+}, the protein undergoes a conformational change, decomposing into apoaequorin

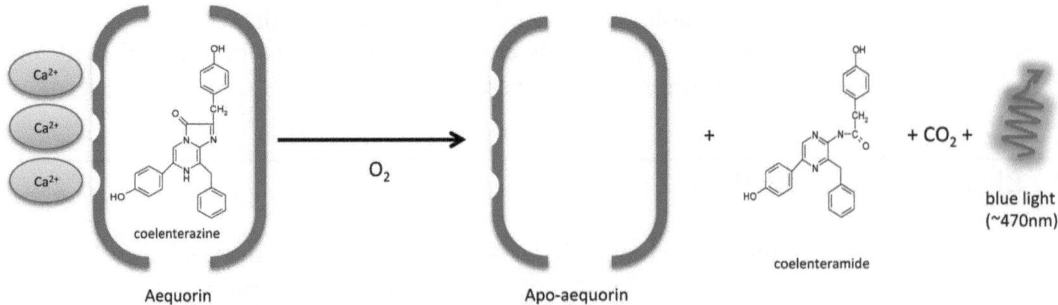

Fig. 1 Schematic representation of the aequorin bioluminescence reaction. When Ca^{2+} binds to aequorin, conformational change of the apoaequorin protein and oxidation of coelenterazine results in blue light emission

while oxidizing coelenterazine into coelenteramide and CO_2 with emission of blue light at 469 nm [19] (Fig. 1).

Typically, the recombinant aequorin is expressed in the cell of interest. When coelenterazine is added to the cells, it will passively diffuse and generate bioluminescence light relative to the total intracellular Ca^{2+} levels available for the reaction. While the high sensitivity of aequorin for Ca^{2+} make this reporter an ideal Ca^{2+} sensor, signal acquisition can be a daunting task due to the protein low light quantum yield (number of photons emitted per protein) combined with the low protein stability [20].

2.2 Luciferases (See Table 1)

2.2.1 Bacterial Luciferases (Lux)

This particular group of luciferases uses the reduced riboflavin phosphate ($FMNH_2$) as their substrate, in addition to a long-chain fatty aldehyde and oxygen. $FMNH_2$ is oxidized to emit a blue–green light at 490 nm [21]. The synergistic expression of all five genetic components of the *luxCDABE* operon produces an autonomous bioluminescence reaction [22]. This property represents a major advantage for lux reporters since light is generated without the need for substrate administration or experimental manipulation. The *lux*A and *lux*B genes encode the α- and β-subunit, respectively, which form the heterodimeric luciferase. The *lux*CDE genes are required for the regeneration of the long-chain fatty aldehyde [22]. The additional components for bacterial luminescence include oxygen and $FMNH_2$, readily available within the cell (Fig. 2). The lux reporters are commonly expressed in bacterial hosts as a means to monitor bacterial growth. Recently, Close et al. described a codon-optimized lux cassette that generates an autonomous bioluminescent system for mammalian expression [23]. This reporter allows whole animal imaging while eliminating the need for substrate administration, a clear advantage over commonly used luciferases such as Firefly, *Gaussia*, and *Renilla* (*see below*). However, the signal intensity remains significantly low as compared to these luciferases. An increase in aldehyde production, which would substantially increase the signal intensity, is hindered by the cytotoxicity of this organic compound [24].

Table 1
Comparison of different luciferases

Luciferase	Origin	Size (kDa)	Substrate	Cofactors	In vitro sensitivity	In vivo sensitivity[a,b]	Secreted	Peak emission (nm)
Bacterial luciferase	Various photo-bacterium species	α subunit: 40 β subunit: 37	$FMNH_2$	O_2, NADPH	+	+	No	490
Firefly	Photinus pyralis	61	D-luciferin	O_2, ATP, Mg^{2+}	+++	++++	No	562
Renilla	Renilla reniformis	36	Coelenterazine	O_2	++	++	No	480
Gaussia	Gaussia princeps	19.9	Coelenterazine	O_2	++++	+++	Yes	480
Cypridina (Vargula)	Vargula hilgendorfi	62	Vargulin	O_2	+++	+++	Yes	460
Metridia	Metridia longa	24	Coelenterazine	O_2	+++	+	Yes	480

[a]As measured with a CCD camera and not through ex vivo blood assays as it can be the case for secreted luciferases
[b]While peak emission is a major determining factor for in vivo sensitivity, other factors such as substrate concentration, administration route, and solubility are also important and can significantly affect the sensitivity of any given luciferase

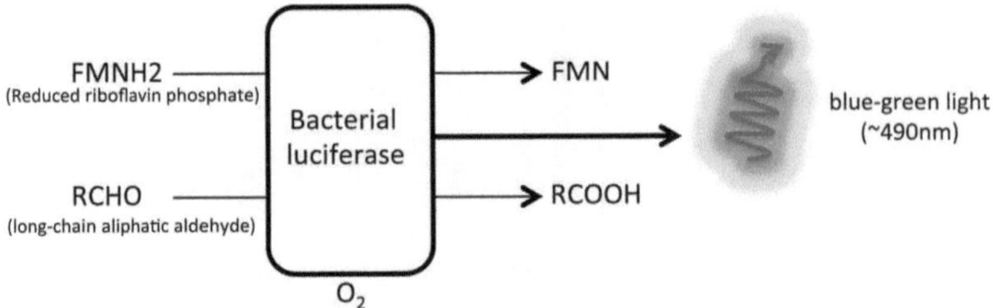

Fig. 2 Schematic representation of the bacterial luciferase reaction. The bacterial luciferase reacts with the reduced riboflavin phosphate in the presence of a long-chain fatty aldehyde to produce a blue–green light

2.2.2 Firefly Luciferase (Fluc)

A monomeric protein (61 kDa) found in the light-emitting organ within the abdomen of the American firefly *Photinus pyralis* [25]. Fluc is the most studied luciferase due to its high quantum yield (originally thought to be around 88 %, however, a more advanced study using a CCD-spectrometer system showed it to be closer to 41 % [26]). This luciferase requires ATP and Mg^{2+} as cofactors in combination with its substrate, beetle D-luciferin (a benzothiazole) [1, 27, 28]. Fluc catalyzes a glow-type bioluminescence reaction generating yellow–green light with a peak emission at 562 nm. For Fluc in vivo BLI, D-luciferin injected intraperitoneally (i.p.) or intravenously (i.v.) has a high biodistribution since it can cross the blood- and placental barriers. Maximum light emission is achieved at 10–12 min after i.p. luciferin injection followed by a slow decay over 60 min [29].

2.2.3 Renilla Luciferase (Rluc)

A monomeric protein (36 kDa) from the sea pansy *Renilla reniformis.* Rluc catalyzes the oxidative decarboxylation of its substrate coelenterazine while emitting blue light with a peak at 480 nm. Like other coelenterates, Rluc generates a flash-type bioluminescence reaction and does not require ATP for activity. A disadvantage of Rluc over Fluc is its low enzymatic turnover and quantum yield (6 %) [30, 31]. Also, the blue emission as well as the poor biodistribution of coelenterazine makes Rluc a less desirable reporter for in vivo imaging [30].

2.2.4 Gaussia Luciferase (Gluc)

A monomeric protein (19.9 kDa) from the marine copepod *Gaussia princeps,* which uses coelenterazine as a substrate. Gluc is the smallest known luciferase, it is naturally secreted and emits a flash light at a peak of 480 nm with a broad emission spectrum extending to 600 nm [32]. Gluc has several advantages over other luciferases: it is over 2000-fold more sensitive than Fluc or Rluc in reporting from mammalian cells and gives a much stronger bioluminescent signal in vivo [32]; Since it is naturally secreted, it can be detected in the conditioned medium of cells expressing this reporter in cell-based assays, and in the blood or other bodily fluids in small animals [33]. On the other hand and similar to Rluc, the blue light emission and the stability of coelenterazine, makes Gluc less favorable for in vivo BLI.

Other luciferases had been described in the literature. Metridia luciferase, a 24 kDa protein with a peak emission at 480 nm, also utilizes coelenterazine as substrate [34]. Cypridina luciferase (also known as Vargula; Vluc), a 62 kDa protein, has a peak emission at 460 nm and utilizes Cypridina luciferin (vargulin) as substrate [35]. Both of these marine luciferases are naturally secreted and do not require ATP for activity.

2.3 Multiplexing BLI Reporters

Despite the advantages of BL reporters over their fluorescent counterparts, the versatility when it comes to different light-emitting spectra remains a downside. Various mutations in the GFP and other fluorescent proteins resulted in a large palette of colored reporters [36]. Once combined together, these reporter variants can be used to monitor various processes within the same experimental sample. Similar mutation studies, to generate luciferases with various light-emission properties or better stability, are now starting to emerge. Branchini et al. generated a green and red variant of Fluc allowing monitoring of two-different activities simultaneously in the same biological sample [37]. A red-shifted variant of Rluc with a peak emission at 547 nm was shown to be better suited for small animal imaging [38]. Recently, new Gluc variants have been characterized with a glow-type bioluminescence reaction, suited for high-throughput functional screening applications [39, 40]. While the spectral diversity of a given luciferase makes it an attractive tool for a multiplex BL assay, the chemical properties of these reporters are also of high relevance. The type of the BL reaction (flash versus glow) and the different substrate chemistry can be exploited for sequential reading of the different luciferase activities. For example, Rluc (or Gluc) luciferase can be combined with Fluc to measure two-different readouts after addition of coelenterazine and D-luciferin, respectively. Recently, Maguire et al. optimized a triple-imaging platform by combining Fluc, Gluc, and Vluc. Their approach allowed monitoring of gene delivery, tumor size, as well as transcription factor activity within the same animal [41]. Such multiplexing approaches are highly valuable for cell-based high-throughput screening, allowing measurement of various parameters within the same biological sample and gathering a maximum amount of readouts in a time- and cost-efficient fashion.

3 BLI Applications

Bioluminescent organisms often use light-emission properties to interact and communicate with their environment and the surroundings organisms. By reproducing those same BL properties,

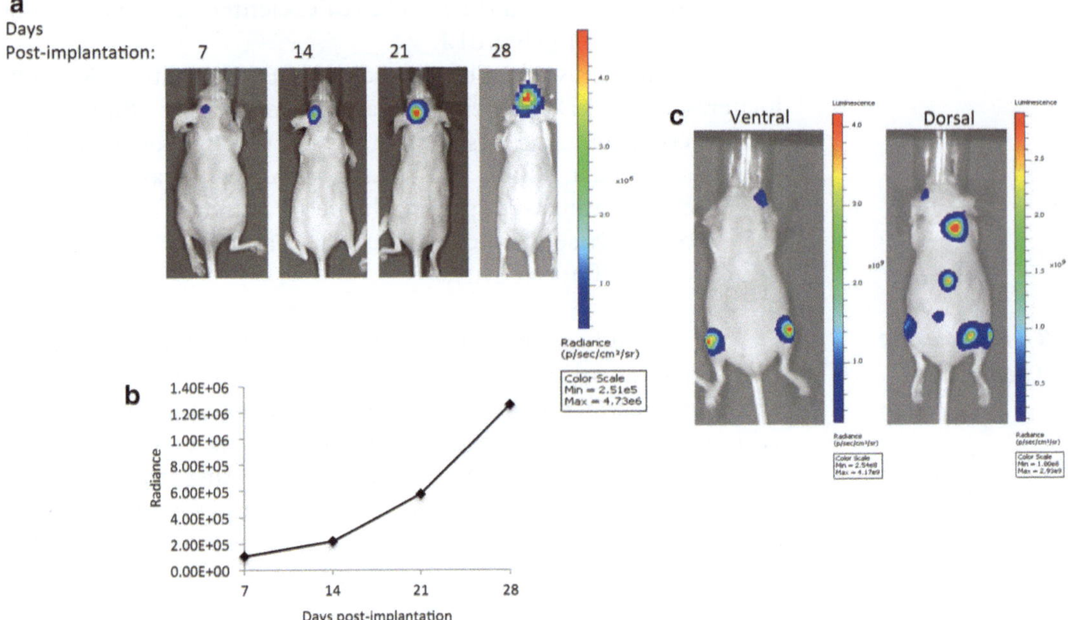

Fig. 3 (**a**, **b**) Monitoring of intracranial tumor growth using bioluminescence imaging. The human glioblastoma cell line U87-MG-Fluc cells (1×10^5 cells) were stereotactically injected into the left midstriatum of nude mice. Animals were imaged for Fluc at different time points. Representative images from the CCD imaging are shown in (**a**). Quantification of the BL signal is represented as radiance (photons/s/cm^2/sr) in (**b**). (**c**) Bioluminescence imaging of a breast cancer metastatic model using a CCD camera. MDA-MB-231-Fluc cells (2×10^5 cells) were implanted intracardially into the animal. Dorsal and ventral images 4 weeks after tumor cells implantation show metastatic lesions in various organs

researchers are able to study similar interactions among cells, genes, and other entities of relevance to the biomedical field.

Several Bioluminescence imaging strategies have been developed to study different cellular and molecular events in living organisms, while providing a spatial and temporal resolution. These imaging paradigms rely on the expression of a foreign protein (luciferase) that is not usually expressed in the cell or the organism of interest. Upon exposure of the luciferase enzyme to the corresponding substrate, light emission can be detected using a luminometer (in cultured cells) or a cooled charge-coupled device (CCD) camera in animal models. Luciferase reporters are commonly used for tracking of cell viability, quantifying gene transfer efficiency, estimating tumor burden or detecting metastatic lesions in animals (Fig. 3). In such studies, the luciferase, expressed under a particular promoter, is introduced to the cell of interest either by plasmid transfection or viral transduction (for stable expression).

BLI has also been used for studying different biological and molecular processes such as cell signaling, transcriptional promoters, gene expression, protein–protein interactions, protein conformational changes, enzymatic activities, protein secretion, and visualization of subcellular proteins. For in vitro assays, BL allows the study of different processes at the nuclear, cytoplasmic, and cellular level as well as cell-free based analysis [42]. One major advantage of this imaging technique is its noninvasive nature allowing the study of biological processes in intact living cells or animals. When using a secreted luciferase, it is possible to longitudinally monitor various biological parameters form the same sample by assaying few microliters of the cell culture medium. These reporters also have major advantages for in vivo studies since they can be detected in bodily fluids such as blood, serum, or urine [43]. Due to its high sensitivity, the Gluc secreted reporter has been used to study tumor growth and therapy, viral replication, viability of circulating stem cells as well as metastatic tumors by directly assaying its activity in the blood of experimental animals [33, 44]. While typically in vivo imaging experiments are performed using small animals models (mice, rats), secreted luciferases can be applied for imaging larger animal models. Gluc has already been used as an ex vivo blood reporter to study gene transfer in the lungs of mice and sheep [45]. Mluc and Vluc can also be used as secreted blood reporters. Mluc, however, is inactivated by serum, thus limiting its use as a secreted blood reporter [46, 47].

4 Drawbacks and Limitations

Just like any other imaging modality, BLI has its advantages and its shortcomings. Oftentimes, data interpretation is performed under the assumption that the luciferase activity is directly correlated to the transcriptional activity of the reporter gene, therefore the BL signal is linear in respect to cell number. However, both endogenous and exogenous factors can impact the various components of the luciferase reaction and might lead to erroneous readouts. It is however important to keep in mind that, based on the assay design, luciferin, luciferases, or cofactors such as ATP can be the variable component for the BL reaction [48]. Some of common problems encountered when using luciferase-based reporters are discussed below. These problems can influence the BL signal generated and should be taken into account when designing a BLI experiment.

- *Signal quantification*: It is difficult to standardize in vitro BL assays since the relative light units (RLUs) measured from a luciferase reaction are arbitrary units. The RLU varies largely from one luminometer or photon detector to the other. For example, interlaboratory BL assays using the same luciferase and luciferin but a different luminometer would most likely yield RLUs with great variability. It is very important to keep in mind that BL is rather a semiquantitative method; its sensitivity depends partly on the luminometer when applied in vitro or the CCD camera for in vivo imaging.

- *Background control*: While the high sensitivity of BLI is partly due to its low background noise, it is still recommended to include a proper background control for your assay. For example, when preforming an in vitro BL assay from cellular lysates or cell supernatant, it is important to use the same buffers or media from cells that do not express the luciferase. The serum among other supplements added to cell culture media can impact the BL readings and must be accounted for. Autoluminescence or light emission by substrates such as coelenterazine, in the absence of the enzyme, can also increase the luminescence background and cause variability.
- *Half-life of the enzyme and stability*: these factors can vary greatly between luciferases and can range from few hours to several days. It is important to determine a priori the half-life of your luciferase based on your assay conditions. Both endogenous and exogenous factors can also affect the luciferase stability. In one example, observations made by Czupryna et al. suggested that the Fluc activity is rapidly lost in apoptotic cells due to oxidative stress and particularly hydrogen peroxide which inhibited the BL signal [49]. The different components of the extracellular medium (cell growth medium, blood, urine, etc.) can also significantly affect the enzyme stability notably for secreted luciferases.
- *Cellular environment*: Both secreted and non-secreted luciferases can be subject to different intra or extracellular conditions which could directly affect their activity. Proteolytic degradation of the enzyme, pH, temperature, and H_2O_2 levels are among many factors that could impact the BL signal. Further, these conditions could indirectly affect the luciferase-expressing cells and impact the proper synthesis, folding, maturation, or secretion of the enzyme.
- *Oxygen, hypoxia, and oxidative stress*: oxygen is a limiting factor for all luciferase reactions. BL assays cannot be conducted under anaerobic conditions, and light emission from hypoxic tissues such as the bulk of a tumor is virtually absent. The lack of oxygen could also indirectly impact the BL reaction by affecting other cofactors needed for certain luciferase reaction (e.g., ATP for Fluc). Moriyama et al. attributed the decrease in Fluc BL signal in hypoxic cells in vitro to an intracellular ATP depletion [50]. In contrast, the Czupryna et al. study attributed this signal decrease in apoptotic cells to hydrogen peroxide [49]. Interestingly, ATP depletion is commonly observed during apoptosis.
- *Substrate availability*: This factor is almost irrelevant for in vitro assays where the substrate and the cofactors needed for

the BL reaction are always in excess to the enzyme. In such settings, light emission is directly proportional to the luciferase concentration. However, substrate availability to the luciferase-expressing cells is critical for in vivo imaging. To generate a strong in vivo photon emission, a sufficient amount of substrate should reach the luciferase-expressing cells and (in the case of non-secreted luciferases) should be taken up by these cells. In this setting, the location of the luciferase-expressing cells and their membrane permeability to the substrate, as well as the amount of substrate injected into the animal can impact the BL signal. Larger tumors have higher substrate uptake compared to smaller size, which could lead to inadequately higher signals [42]. Finally, whether the luciferin is metabolized or directly cleared via the kidneys depends on the type of substrate used and on various physiological parameters such as temperature, heart rate, and breathing of the animal. Most frequently, animals are under deep anesthesia during imaging and maintaining similar physiological condition among the different subjects is important to minimize artifacts due to substrate availability.

- *Luciferin efflux*: The luciferase substrate can be actively pumped outside of the cells thus reducing the BL signal. The ATP-binding cassette superfamily of multidrug efflux pumps is notorious for conferring chemoresistance in tumor cells. Two of these efflux pumps, the ABCG2 (BCRP) and ABCB1/Pgp, are substrates for D-luciferin and coelenterazine, respectively [51, 52]. Their expression can significantly reduce the Fluc or Rluc/Gluc signal. Compounds that affect membrane transport proteins could significantly impact the BL signal.
- *Administration route of the substrate*: this factor is very critical for in vivo imaging. ^{14}C-labeled D-luciferin showed different biodistribution after i.p. or i.v. injections [53]. When the substrate was administered i.v., a more homogeneous distribution among tissues could be obtained as compared to i.p administration, while the latter yielded a prolonged organ uptake of D-luciferin. When using luciferases with glow-type reaction kinetics, such as Fluc, an incubation step of few minutes following i.p. substrate administration (5–15 min) would allow its absorption through the peritoneum and a better distribution throughout different tissues. For luciferases with flash-type kinetics (such as Gluc and Rluc), it is best to image immediately after i.v. substrate administration, to ensure the highest signal intensity. In addition to the reaction kinetics, other factors should be considered when deciding on the administration route including the amount of substrate injected, solubility of the substrate, the solvent used, as well as the half-life of the substrate in vivo. The i.v. route is more

appropriate when a small amount of substrate is administered or when the substrate is rapidly cleared, as is the case for coelenterazine.

- *Promoter activity*: Bioluminescent reporters are commonly used to study transcriptional activity. *Cis*-transcriptional reporter systems allow the analysis of gene expression and gene regulation. This is performed by either generating point mutations/deletions in a promoter region of a gene of interest or the use of different transcription factor binding sites linked to a minimal promoter to drive the expression of a luciferase [54–57]. This approach is useful for reporting different events that affect transgene expression such as signal transduction [58, 59], receptor activation [60, 61], and transcription factors activity [62–66], thus complementing conventional in vitro methods of molecular biology and biochemistry. However, exogenous factors can impact the promoter activity. For example, the cytomegalovirus (CMV) promoter, commonly used to express the luciferase reporter, can be upregulated by different chemotherapeutic agents or other extrinsic stimuli [67–69]. This could lead to an underestimation of the efficacy of a given drug treatment. The SV40 promoter/enhancer is less prone to variation induced by common chemotherapeutics or irradiation and therefore is a better choice. Both CMV and the SV40 promoter/enhancers activities can also be negatively affected by exogenous treatments such as interferon-gamma [70]. Depending on the assay conditions, it is important to determine the adequate promoter for your experiment.
- *Light quenching and scattering*: bioluminescence imaging in deep tissues is not trivial due to light absorption by pigmented molecules (e.g., hemoglobin and melanin) and light scattering by mammalian tissues. Highly vascularized organs emit lower light signals compared to skin or muscle [71]. Highly pigmented mice also yield lower light output due to melanin light absorption; the signal is lower in black mice compared to white mice for example [72]. Hair and fur can also scatter and attenuate light signal. This problem could be easily overcome by removing the animal's hair through depilation or shaving. However complications may arise since hair removal can disrupt normal hair growth cycle and subsequently change the skin pigmentation [73]. Another option is to use mice lacking a fur coat. The emergence of red-shifted luciferases helps overcome such problems and greatly enhances the sensitivity of in vivo BLI in deep tissues [37, 74, 75].
- *Spatial resolution*: another consequence of light scattering is the low spatial resolution of BLI (1–2 mm) [71]. This low resolution yields a poor localization of the BL source emanating from dispersed tumor cells and small metastatic niches for instance,

within the imaging subject. This limitation is becoming less problematic with the emergence of multi-modal imaging and bioluminescence tomography capable of three-dimensional image reconstruction and acquiring the signal at multiple angles.

- *Luciferase modulators*: Various synthetic or natural small compounds could affect the luciferase activity and impact the BL signal. The modulators in question affect the enzyme itself and not its transcription or translation. Since Fluc is the most widely used luciferase, most studies have focused on compounds that affect its activity. Such compounds can directly interact with the reporter leading to an increase, decrease, or a bell-shaped concentration response curve of the emitted BL signal [76]. This bell-shaped curve occurs when the compound at low concentrations stabilizes the Fluc enzyme leading to its apparent activation, while at higher concentration an inhibition of the luciferase activity is observed [76]. A hypothetical increase in luciferase half-life by 30 % can lead to 150 % increase in BL signal within 12 h [77]. Numerous studies have reported the identification of small-molecule inhibitors of Fluc [76, 78–80]. Inhibitors of other luciferases have not yet been identified; however, it is safe to assume that various small molecules are also capable of affecting the activities of these reporters. In the case of secreted luciferases such as Gaussia or Cypridina luciferase, a different type of modulators that affect protein secretion or proper folding, could also significantly impact the luciferase signal [81]. These problems could lead to deceptive "hit identification" when luciferase-based high-throughput assays are employed and large compound libraries are screened. Scrupulous data analysis and additional validation steps using different assays (preferably not luciferase-based) can help eliminate any compound/luciferase-based artifact.
- Some anesthetics have been shown to bind to Fluc and inhibit its activity in vitro; however, they only showed moderate inhibitory effect on BL signal during in vivo imaging [82]. Standardizing dosage and time exposure of animals to anesthetics would ensure more reproducible results.
- *Single cell analysis*: Similar to many other molecular imaging techniques, BLI measures an average across a population of cells, thereby decreasing the sensitivity to phenotypic changes in a certain subpopulation [83]. Unlike techniques such as flow cytometry or microscopy, BL reports from the population as a whole and not from individual cells. For applications that require imaging of individual cells, a high-intensity signal is required and therefore fluorescence is the reporter of choice. In addition to its low sensitivity in reporting from single cells, cellular expression of the luciferase might vary between different subpopulations. This is generally due to different copy

numbers of the luciferase gene or variation in integration sites when a virus is used for stable luciferase expression. This variation might increase under culture conditions where a certain subpopulation of cells might have a survival advantage.

5 Conclusions

As the BLI field continues to grow, more robust and accurate imaging tools are still emerging and efforts to overcome various technical challenges are ongoing. Significant improvements targeting the three pivotal variables impacting the BL reaction: the enzyme itself, the substrate and the photon detectors have already been made. Enzymes with higher sensitivity, deeper light penetrance and better stability have been generated. Substrates with better sensitivity, stability, solubility and biodistribution are now commercially available. Finally light-detectors with enhanced sensitivity and the ability to reconstruct three-dimensional images are starting to emerge, albeit rather slowly and with a significant cost. Choosing the right luciferase that is best suited for a particular application could also help overcome many of the hurdles discussed above. The light emission of the luciferase, the type of light generated, the substrate required, the promoter driving the enzyme expression and many other factors should be taken into consideration. Validating the BL results using orthogonal assays and critical appraisal of the data obtained is also a great way to avoid erroneous interpretations of results and ensure strong reproducible results. In conclusion, the versatility of BLI offers limitless applications for biomedical investigators. A hands-on experience as well as an extensive knowledge of the advantages and limitations of this technique could help overcome most problems and turn some into assets.

Acknowledgments

This work was supported by a Fellowship from the American Brain Tumor Association. The author is grateful for Dr. Bakhos A. Tannous for his suggestions and his critical reading of this manuscript.

References

1. Massoud TF, Gambhir SS (2003) Molecular imaging in living subjects: seeing fundamental biological processes in a new light. Genes Dev 17(5):545–580
2. Weissleder R, Pittet MJ (2008) Imaging in the era of molecular oncology. Nature 452(7187): 580–589
3. Contag CH, Bachmann MH (2002) Advances in in vivo bioluminescence imaging of gene expression. Annu Rev Biomed Eng 4: 235–260
4. O'Brien MA, Daily WJ, Hesselberth PE, Moravec RA, Scurria MA, Klaubert DH, Bulleit RF, Wood KV (2005) Homogeneous,

bioluminescent protease assays: caspase-3 as a model. J Biomol Screen 10(2):137–148

5. Ketteler R, Sun Z, Kovacs KF, He WW, Seed B (2008) A pathway sensor for genome-wide screens of intracellular proteolytic cleavage. Genome Biol 9(4):R64
6. Mizushima N, Kuma A (2008) Autophagosomes in GFP-LC3 transgenic mice. Methods Mol Biol 445:119–124
7. Azadniv M, Dugger K, Bowers WJ, Weaver C, Crispe IN (2007) Imaging CD8+ T cell dynamics in vivo using a transgenic luciferase reporter. Int Immunol 19(10):1165–1173. doi:dxm086 [pii] 10.1093/intimm/dxm086
8. Baumjohann D, Lutz MB (2006) Non-invasive imaging of dendritic cell migration in vivo. Immunobiology 211(6–8):587–597
9. Nishijo K, Hosoyama T, Bjornson CR, Schaffer BS, Prajapati SI, Bahadur AN, Hansen MS, Blandford MC, McCleish AT, Rubin BP, Epstein JA, Rando TA, Capecchi MR, Keller C (2009) Biomarker system for studying muscle, stem cells, and cancer in vivo. FASEB J 23(8):2681–2690. doi:fj.08-128116 [pii] 10.1096/fj.08-128116
10. Shah K, Bureau E, Kim DE, Yang K, Tang Y, Weissleder R, Breakefield XO (2005) Glioma therapy and real-time imaging of neural precursor cell migration and tumor regression. Ann Neurol 57(1):34–41. doi:10.1002/ana.20306
11. Zhao H, Tang C, Cui K, Ang BT, Wong ST (2009) A screening platform for glioma growth and invasion using bioluminescence imaging. Laboratory investigation. J Neurosurg 111(2):238–246. doi:10.3171/2008.8.JNS08644 10.3171/2008.8.JNS08644 [pii]
12. Livingston JN (1999) Genetically engineered mice in drug development. J Intern Med 245(6):627–635
13. Iyer M, Sato M, Johnson M, Gambhir SS, Wu L (2005) Applications of molecular imaging in cancer gene therapy. Curr Gene Ther 5(6):607–618
14. Rome C, Couillaud F, Moonen CT (2007) Gene expression and gene therapy imaging. Eur Radiol 17(2):305–319
15. Shimomura O (1985) Bioluminescence in the sea: photoprotein systems. Symp Soc Exp Biol 39:351–372
16. Johnson FH, Shimomura O (1972) Enzymatic and nonenzymatic bioluminescence. Photophysiology 7:275–334
17. Shimomura O, Johnson FH, Saiga Y (1962) Extraction, purification and properties of aequorin, a bioluminescent protein from the luminous hydromedusan, Aequorea. J Cell Comp Physiol 59:223–239
18. Head JF, Inouye S, Teranishi K, Shimomura O (2000) The crystal structure of the photoprotein aequorin at 2.3 A resolution. Nature 405(6784):372–376. doi:10.1038/35012659
19. Shimomura O, Johnson FH, Morise H (1974) Mechanism of the luminescent intramolecular reaction of aequorin. Biochemistry 13(16):3278–3286
20. Baubet V, Le Mouellic H, Campbell AK, Lucas-Meunier E, Fossier P, Brulet P (2000) Chimeric green fluorescent protein-aequorin as bioluminescent Ca2+ reporters at the single-cell level. Proc Natl Acad Sci U S A 97(13):7260–7265
21. Meighen EA (1994) Genetics of bacterial bioluminescence. Annu Rev Genet 28:117–139. doi:10.1146/annurev.ge.28.120194.001001
22. Meighen EA (1991) Molecular biology of bacterial bioluminescence. Microbiol Rev 55(1):123–142
23. Close DM, Patterson SS, Ripp S, Baek SJ, Sanseverino J, Sayler GS (2010) Autonomous bioluminescent expression of the bacterial luciferase gene cassette (lux) in a mammalian cell line. PloS One 5(8):e12441. doi:10.1371/journal.pone.0012441
24. Hollis RP, Lagido C, Pettitt J, Porter AJ, Killham K, Paton GI, Glover LA (2001) Toxicity of the bacterial luciferase substrate, n-decyl aldehyde, to Saccharomyces cerevisiae and Caenorhabditis elegans. FEBS Lett 506(2):140–142
25. de Wet JR, Wood KV, DeLuca M, Helinski DR, Subramani S (1987) Firefly luciferase gene: structure and expression in mammalian cells. Mol Cell Biol 7(2):725–737
26. Ando YNK, Yamada N, Enomoto T, Irie T, Kubota H, Ohmiya Y, Akiyama H (2008) Firefly bioluminescence quantum yield and colour change by pH-sensitive green emission. Nat Photonics 2:44–47
27. de Wet JR, Wood KV, Helinski DR, DeLuca M (1985) Cloning of firefly luciferase cDNA and the expression of active luciferase in Escherichia coli. Proc Natl Acad Sci U S A 82(23):7870–7873
28. Lembert N, Idahl LA (1995) Regulatory effects of ATP and luciferin on firefly luciferase activity. Biochem J 305(Pt 3):929–933
29. Paroo Z, Bollinger RA, Braasch DA, Richer E, Corey DR, Antich PP, Mason RP (2004) Validating bioluminescence imaging as a high-throughput, quantitative modality for assessing tumor burden. Mol Imaging 3(2):117–124
30. Bhaumik S, Gambhir SS (2002) Optical imaging of Renilla luciferase reporter gene expression in living mice. Proc Natl Acad Sci U S A 99(1):377–382

31. Matthews JC, Hori K, Cormier MJ (1977) Purification and properties of Renilla reniformis luciferase. Biochemistry 16(1):85–91
32. Tannous BA, Kim DE, Fernandez JL, Weissleder R, Breakefield XO (2005) Codon-optimized Gaussia luciferase cDNA for mammalian gene expression in culture and in vivo. Mol Ther 11(3):435–443
33. Wurdinger T, Badr C, Pike L, de Kleine R, Weissleder R, Breakefield XO, Tannous BA (2008) A secreted luciferase for ex vivo monitoring of in vivo processes. Nat Methods 5(2):171–173. doi:10.1038/nmeth.1177
34. Markova SV, Golz S, Frank LA, Kalthof B, Vysotski ES (2004) Cloning and expression of cDNA for a luciferase from the marine copepod Metridia longa. A novel secreted bioluminescent reporter enzyme. J Biol Chem 279(5): 3212–3217. doi:10.1074/jbc.M309639200
35. Thompson EM, Nagata S, Tsuji FI (1989) Cloning and expression of cDNA for the luciferase from the marine ostracod Vargula hilgendorfii. Proc Natl Acad Sci U S A 86(17): 6567–6571
36. Shaner NC, Steinbach PA, Tsien RY (2005) A guide to choosing fluorescent proteins. Nat Methods 2(12):905–909. doi:10.1038/nmeth819
37. Branchini BR, Ablamsky DM, Murtiashaw MH, Uzasci L, Fraga H, Southworth TL (2007) Thermostable red and green light-producing firefly luciferase mutants for bioluminescent reporter applications. Anal Biochem 361(2): 253–262. doi:10.1016/j.ab.2006.10.043
38. Loening AM, Wu AM, Gambhir SS (2007) Red-shifted Renilla reniformis luciferase variants for imaging in living subjects. Nat Methods 4(8):641–643
39. Degeling MH, Maguire CA, Bovenberg MS, Tannous BA (2012) Sensitive assay for mycoplasma detection in mammalian cell culture. Anal Chem 84(9):4227–4232. doi:10.1021/ac2033112
40. Maguire CA, Deliolanis NC, Pike L, Niers JM, Tjon-Kon-Fat LA, Sena-Esteves M, Tannous BA (2009) Gaussia luciferase variant for high-throughput functional screening applications. Anal Chem 81(16):7102–7106. doi:10.1021/ac901234r
41. Maguire CA, Bovenberg MS, Crommentuijn MH, Niers JM, Kerami M, Teng J, Sena-Esteves M, Badr CE, Tannous BA (2013) Triple bioluminescence imaging for in vivo monitoring of cellular processes. Mol Ther Nucleic Acids (in press)
42. Badr CE, Tannous BA (2011) Bioluminescence imaging: progress and applications. Trends Biotechnol 29(12):624–633. doi:10.1016/j.tibtech.2011.06.010
43. Tannous BA, Teng J (2011) Secreted blood reporters: insights and applications. Biotechnol Adv 29(6):997–1003. doi:10.1016/j.biotechadv.2011.08.021
44. Chung E, Yamashita H, Au P, Tannous BA, Fukumura D, Jain RK (2009) Secreted Gaussia luciferase as a biomarker for monitoring tumor progression and treatment response of systemic metastases. PloS One 4(12):e8316. doi:10.1371/journal.pone.0008316
45. Griesenbach U, Vicente CC, Roberts MJ, Meng C, Soussi S, Xenariou S, Tennant P, Baker A, Baker E, Gordon C, Vrettou C, McCormick D, Coles R, Green AM, Lawton AE, Sumner-Jones SG, Cheng SH, Scheule RK, Hyde SC, Gill DR, Collie DD, McLachlan G, Alton EW (2011) Secreted Gaussia luciferase as a sensitive reporter gene for in vivo and ex vivo studies of airway gene transfer. Biomaterials 32(10):2614–2624. doi:10.1016/j.biomaterials.2010.12.001
46. Lupold SE, Johnson T, Chowdhury WH, Rodriguez R (2012) A real time Metridia luciferase based non-invasive reporter assay of mammalian cell viability and cytotoxicity via the beta-actin promoter and enhancer. PloS One 7(5):e36535. doi:10.1371/journal.pone.0036535
47. Hiramatsu N, Kasai A, Meng Y, Hayakawa K, Yao J, Kitamura M (2005) Alkaline phosphatase vs luciferase as secreted reporter molecules in vivo. Anal Biochem 339(2):249–256. doi:10.1016/j.ab.2005.01.023
48. Fan F, Wood KV (2007) Bioluminescent assays for high-throughput screening. Assay Drug Dev Technol 5(1):127–136. doi:10.1089/adt.2006.053
49. Czupryna J, Tsourkas A (2011) Firefly luciferase and RLuc8 exhibit differential sensitivity to oxidative stress in apoptotic cells. PloS One 6(5):e20073. doi:10.1371/journal.pone.0020073
50. Moriyama EH, Niedre MJ, Jarvi MT, Mocanu JD, Moriyama Y, Subarsky P, Li B, Lilge LD, Wilson BC (2008) The influence of hypoxia on bioluminescence in luciferase-transfected gliosarcoma tumor cells in vitro. Photochem Photobiol Sci 7(6):675–680. doi:10.1039/b719231b
51. Zhang Y, Bressler JP, Neal J, Lal B, Bhang HE, Laterra J, Pomper MG (2007) ABCG2/BCRP expression modulates D-Luciferin based bioluminescence imaging. Cancer Res 67(19):9389–9397. doi:10.1158/0008-5472.CAN-07-0944
52. Pichler A, Prior JL, Piwnica-Worms D (2004) Imaging reversal of multidrug resistance in living

mice with bioluminescence: MDR1 P-glycoprotein transports coelenterazine. Proc Natl Acad Sci U S A 101(6):1702–1707. doi:10.1073/pnas.0304326101

53. Berger F, Paulmurugan R, Bhaumik S, Gambhir SS (2008) Uptake kinetics and biodistribution of 14C-D-luciferin—a radiolabeled substrate for the firefly luciferase catalyzed bioluminescence reaction: impact on bioluminescence based reporter gene imaging. Eur J Nucl Med Mol Imaging 35(12):2275–2285. doi: 10.1007/s00259-008-0870-6
54. Lin MC, Li JJ, Wang EJ, Princler GL, Kauffman FC, Kung HF (1997) Ethanol down-regulates the transcription of microsomal triglyceride transfer protein gene. Faseb J 11(13): 1145–1152
55. Stratowa C, Audette M (1995) Transcriptional regulation of the human intercellular adhesion molecule-1 gene: a short overview. Immunobiology 193(2–4):293–304
56. Ueda T, Akiyama N, Sai H, Oya N, Noda M, Hiraoka M, Kizaka-Kondoh S (2001) c-IAP2 is induced by ionizing radiation through NF-kappaB binding sites. FEBS Lett 491(1–2): 40–44
57. Wrana JL, Attisano L, Wieser R, Ventura F, Massague J (1994) Mechanism of activation of the TGF-beta receptor. Nature 370(6488): 341–347
58. Herbst KJ, Allen MD, Zhang J (2009) The cAMP-dependent protein kinase inhibitor H-89 attenuates the bioluminescence signal produced by Renilla Luciferase. PloS One 4(5):e5642. doi:10.1371/journal.pone.0005642
59. Zhou S, Zawel L, Lengauer C, Kinzler KW, Vogelstein B (1998) Characterization of human FAST-1, a TGF beta and activin signal transducer. Mol Cell 2(1):121–127
60. Ciana P, Raviscioni M, Mussi P, Vegeto E, Que I, Parker MG, Lowik C, Maggi A (2003) In vivo imaging of transcriptionally active estrogen receptors. Nat Med 9(1):82–86, 10.1038/nm809 nm809 [pii]
61. von Bulow GU, Bram RJ (1997) NF-AT activation induced by a CAML-interacting member of the tumor necrosis factor receptor superfamily. Science 278(5335):138–141
62. Brunet A, Bonni A, Zigmond MJ, Lin MZ, Juo P, Hu LS, Anderson MJ, Arden KC, Blenis J, Greenberg ME (1999) Akt promotes cell survival by phosphorylating and inhibiting a Forkhead transcription factor. Cell 96(6):857–868, S0092-8674(00)80595-4 [pii]
63. Durocher D, Charron F, Warren R, Schwartz RJ, Nemer M (1997) The cardiac transcription factors Nkx2-5 and GATA-4 are mutual cofactors. Embo J 16(18):5687–5696
64. Macian F, Garcia-Rodriguez C, Rao A (2000) Gene expression elicited by NFAT in the presence or absence of cooperative recruitment of Fos and Jun. Embo J 19(17):4783–4795
65. Phippard D, Manning AM (2003) Screening for inhibitors of transcription factors using luciferase reporter gene expression in transfected cells. Methods Mol Biol 225:19–23
66. Tran H, Brunet A, Grenier JM, Datta SR, Fornace AJ Jr, DiStefano PS, Chiang LW, Greenberg ME (2002) DNA repair pathway stimulated by the forkhead transcription factor FOXO3a through the Gadd45 protein. Science 296(5567):530–534
67. Svensson RU, Barnes JM, Rokhlin OW, Cohen MB, Henry MD (2007) Chemotherapeutic agents up-regulate the cytomegalovirus promoter: implications for bioluminescence imaging of tumor response to therapy. Cancer Res67(21):10445–10454.doi:10.1158/0008-5472.CAN-07-1955
68. Badr CE, Niers JM, Tjon-Kon-Fat LA, Noske DP, Wurdinger T, Tannous BA (2009) Real-time monitoring of nuclear factor kappaB activity in cultured cells and in animal models. Mol Imaging 8(5):278–290
69. Hingorani M, White CL, Zaidi S, Merron A, Peerlinck I, Gore ME, Nutting CM, Pandha HS, Melcher AA, Vile RG, Vassaux G, Harrington KJ (2008) Radiation-mediated up-regulation of gene expression from replication-defective adenoviral vectors: implications for sodium iodide symporter gene therapy. Clin Cancer Res 14(15):4915–4924. doi:10.1158/1078-0432.CCR-07-4049
70. Harms JS, Oliveira SC, Splitter GA (1999) Regulation of transgene expression in genetic immunization. Braz J Med Biol Res 32(2): 155–162
71. O'Neill K, Lyons SK, Gallagher WM, Curran KM, Byrne AT (2010) Bioluminescent imaging: a critical tool in pre-clinical oncology research. J Pathol 220(3):317–327. doi:10.1002/path.2656
72. Edinger M, Cao YA, Hornig YS, Jenkins DE, Verneris MR, Bachmann MH, Negrin RS, Contag CH (2002) Advancing animal models of neoplasia through in vivo bioluminescence imaging. Eur J Cancer 38(16):2128–2136
73. Curtis A, Calabro K, Galarneau JR, Bigio IJ, Krucker T (2011) Temporal variations of skin pigmentation in C57BL/6 mice affect optical bioluminescence quantitation. Mol Imaging Biol 13(6):1114–1123. doi:10.1007/s11307-010-0440-8

74. Branchini BR, Southworth TL, Khattak NF, Michelini E, Roda A (2005) Red- and green-emitting firefly luciferase mutants for bioluminescent reporter applications. Anal Biochem 345(1):140–148, S0003-2697(05)00532-4 [pii] 10.1016/j.ab.2005.07.015
75. Caysa H, Jacob R, Muther N, Branchini B, Messerle M, Soling A (2009) A redshifted codon-optimized firefly luciferase is a sensitive reporter for bioluminescence imaging. Photochem Photobiol Sci 8(1):52–56. doi:10.1039/b814566k
76. Thorne N, Shen M, Lea WA, Simeonov A, Lovell S, Auld DS, Inglese J (2012) Firefly luciferase in chemical biology: a compendium of inhibitors, mechanistic evaluation of chemotypes, and suggested use as a reporter. Chem Biol 19(8):1060–1072. doi:10.1016/j.chembiol.2012.07.015
77. Auld DS, Thorne N, Nguyen DT, Inglese J (2008) A specific mechanism for nonspecific activation in reporter-gene assays. ACS Chem Biol 3(8):463–470. doi:10.1021/cb8000793
78. Pang YP, Park JG, Wang S, Vummenthala A, Mishra RK, McLaughlin JE, Di R, Kahn JN, Tumer NE, Janosi L, Davis J, Millard CB (2011) Small-molecule inhibitor leads of ribosome-inactivating proteins developed using the doorstop approach. PloS One 6(3):e17883. doi:10.1371/journal.pone.0017883
79. Auld DS, Lovell S, Thorne N, Lea WA, Maloney DJ, Shen M, Rai G, Battaile KP, Thomas CJ, Simeonov A, Hanzlik RP, Inglese J (2010) Molecular basis for the high-affinity binding and stabilization of firefly luciferase by PTC124. Proc Natl Acad Sci U S A 107(11):4878–4883. doi:10.1073/pnas.0909141107
80. Auld DS, Thorne N, Maguire WF, Inglese J (2009) Mechanism of PTC124 activity in cell-based luciferase assays of nonsense codon suppression. Proc Natl Acad Sci U S A 106(9):3585–3590. doi:10.1073/pnas.0813345106
81. Badr CE, Hewett JW, Breakefield XO, Tannous BA (2007) A highly sensitive assay for monitoring the secretory pathway and ER stress. PloS One 2(6):e571. doi:10.1371/journal.pone.0000571
82. Keyaerts M, Remory I, Caveliers V, Breckpot K, Bos TJ, Poelaert J, Bossuyt A, Lahoutte T (2012) Inhibition of firefly luciferase by general anesthetics: effect on in vitro and in vivo bioluminescence imaging. PloS One 7(1):e30061. doi:10.1371/journal.pone.0030061
83. Feng Y, Mitchison TJ, Bender A, Young DW, Tallarico JA (2009) Multi-parameter phenotypic profiling: using cellular effects to characterize small-molecule compounds. Nat Rev Drug Discov 8(7):567–578

Part I

In Vitro Bioassays

Chapter 2

Extraction and Quantification of Adenosine Triphosphate in Mammalian Tissues and Cells

Junji Chida and Hiroshi Kido

Abstract

Adenosine 5′-triphosphate (ATP) is the "energy currency" of organisms and plays central roles in bioenergetics, whereby its level is used to evaluate cell viability, proliferation, death, and energy transmission. In this chapter, we describe an improved and efficient method for extraction of ATP from tissues and cells using phenol-based reagents. The chaotropic extraction reagents reported so far co-precipitate ATP with insoluble proteins during extraction and with salts during neutralization. In comparison, the phenol-based reagents extract ATP well without the risks of co-precipitation. The extracted ATP can be quantified by the luciferase assay or high-performance liquid chromatography.

Key words ATP measurement, ATP extraction, Luciferase assay, Phenol-based extraction, Nucleotide extraction

1 Introduction

Measurement of adenosine 5′-triphosphate (ATP) levels is widely used to monitor and evaluate energy stasis and metabolic activity in various cells and tissues [1–6]. Depletion of ATP is thus a sensitive marker of impaired cellular function and viability. For quantitative determination of ATP levels, bioluminescence assays using the luciferin–luciferase system are currently the most popular methods due to their high sensitivity and specificity [7]. The initial step for measuring ATP in cells and tissues is extracting the ATP away from the surrounding source material, and the solubilized ATP free from proteins is measured by the luciferase assay. Therefore, the efficiency of ATP extraction is the major determinant of accuracy of the assay.

The ATP extraction media used for bacteria, plants, and mammalian cells and tissues reported so far are boiling water and buffers [8–10], organic solvents [8, 11], acids [8, 10], and proteinase-K [10]. Among the extraction media, many of the manufacturers of commercially available ATP assay kits recommend chaotropic

Christian E. Badr (ed.), *Bioluminescent Imaging: Methods and Protocols*, Methods in Molecular Biology, vol. 1098, DOI 10.1007/978-1-62703-718-1_2,

extraction reagents, such as trichloroacetic acid (TCA), perchloric acid (PCA), and ethylene glycol (EG). However, we recently found that many of these extraction reagents are not suitable for ATP extraction from mammalian tissues, particularly materials with high protein concentrations, because ATP from such tissues is not fully extracted in the homogenization and deproteinization steps, due to co-precipitation of ATP with the insolubilized proteins or adsorption to the acid-salt precipitate during neutralization of the acid extract [12, 13].

In this chapter, we introduce new medium containing phenol-based reagents for optimal ATP extraction from various tissues and cells that can be used even for samples with high protein concentrations. Phenol-based reagents are generally used for DNA and RNA extraction and allow effective and convenient extraction of various nucleotides, including ATP, from tissues and cells. ATP concentration in the extracts was measured by firefly luciferase assay or by nucleotide separation performed by high-performance liquid chromatography (HPLC) on a reverse-phase chromatography column. ATP levels extracted by phenol-based reagents were over 17.8-fold higher than those extracted by TCA [14]. Here we report a simple, rapid, and reliable Tris–EDTA–saturated phenol (phenol–TE) extraction method for ATP measurement in tissues and cells.

2 Materials

All chemicals used are of analytical grade, and all dilutions are made with high-purity deionized water (18 MΩ/cm resistance) obtained from a Milli-Q water purification system (Millipore, Bedford, MA). All solutions are RNase free, and RNase-free glassware and plasticware are used in all experiments. The method used to prepare RNase-free solutions and labware was described previously by Sambrook et al. [15].

2.1 Phenol–TE Reagent

1. Redistilled Phenol, molecular biology grade, ≥99.0 %: Stored in aliquots at −20 °C.
2. 8-hydroxyquinoline.
3. 0.5 M Tris–HCl buffer, pH 8.0.
4. 0.1 M Tris–HCl buffer, pH 8.0.
5. 10 mM Tris–HCl buffer, pH 8.0.
6. 0.5 M ethylenediaminetetraacetic acid disodium salt dihydrate (EDTA), pH 8.0.

2.2 ATP Extraction Medium

1. Phenol–TE reagent for ATP extraction from animal tissues.
2. Chloroform, ≥99.0 %.
3. Polypropylene tube, 14 mL.
4. Ultra-Turrax® homogenizer (Ika Japan, Nara, Japan).

5. Phenol–TE/chloroform/deionized water (6:2:2) solution for ATP extraction from cells.
6. Homogenozation buffer for extraction of ATP from cells and tissues with TCA and EG: 0.25 M sucrose and 10 mM HEPES–NaOH buffer, pH 7.4.
7. TCA, ≥99.0 %: 10 % solution in deionized water.
8. Neutralization buffer for TCA extract: 1 M Tris–acetate buffer, pH 7.75.
9. EG reagent for ATP extraction from cells and tissues (*see* **Note 1**).

2.3 Luciferin–Luciferase Assay

1. ATP stock solution for preparation of ATP standard: 10 mg ATP disodium salt trihydrate, crystallized (>98 % purity, $C_{10}H_{14}N_5Na_2O_{13}P_3 \cdot 3H_2 = 605.19$) is dissolved by addition of 1 mL of high-purity deionized water in the vial. The final concentration of ATP in the stock solution is 16.5 mM (*see* **Note 2**).
2. Luciferin–luciferase reaction buffer: 20 mM Tricine/1 mM Mg-carbonate hydroxide, pentahydrate/2.7 mM $MgSO_4$/33.3 mM Dithiothreitol.
3. Firefly luciferase (luciferase) and its substrate D-luciferin (luciferin) (*see* **Note 3**).
4. Luminometer (TD-20/20: Turner-Designs, Sunnyvale, CA).

2.4 HPLC

1. HPLC (LaChrom D-7000 HPLC System, Hitachi, Tokyo).
2. Reverse-phase chromatography column (TSK-GEL Amide-80, 4.6×250 mm, Tosoh, Tokyo).
3. Mobile phase: 70 % Acetonitrile, 30 % 75 mM KH_2PO_4 (v/v).

3 Methods

3.1 Preparation of Phenol–TE Reagent

1. Place the bottle of phenol under a fume hood and warm to room temperature. Set a water bath to 65–68 °C under the fume hood (*see* **Note 4**).
2. Place the bottle of phenol in the 65–68 °C water bath to melt the phenol crystals. This should take approximately 2 h.
3. Add 100 mg of 8-hydroxyquinoline to 100 mL of phenol (0.1 % 8-hydroxyquinoline) in a glass beaker containing a stirring bar.
4. Gently pour 100 mL of the melted phenol using a RNase-free glass pipette and mix to dissolve the 8-hydroxyquinoline. The phenol will turn yellow due to the production of 8-hydroxyquinoline, which is used as an antioxidant.
5. Add an equal volume (100 mL) of 0.5 M Tris–HCl buffer, pH 8.0, at room temperature and cover the beaker with aluminum foil and stir for 10–15 min with magnetic stirred under the hood.

6. Stop the stirrer and let the two phases separate over approximately 30 min. Then aspirate as much as possible the aqueous layer (top) with portable pipetaid using a 25 mL glass pipette (*see* **Note 5**).
7. Add 400 mL of 0.1 M Tris–HCl buffer, pH 8.0 and mix with magnetic stirrer for 15 min under the hood.
8. Turn off the stirrer and let the two phases separate for approximately 30 min. Aspirate as much as possible of the aqueous layer (top) with portable pipetaid. Leave about 1 cm layer of the buffer over the phenol.
9. Add 400 mL of 0.1 M Tris–HCl buffer, pH 8.0 two more times, for a total of three times, and follow each addition by **steps** 7 and **8**.
10. Check the pH of the phenol phase using pH paper; it should be 8.0 or >7.8. (Do not use a pH meter because phenol will melt the probe!) (*see* **Note 6**).
11. After the adjustment of the pH value, add 200 mL of 10 mM Tris–HCl buffer, pH 8.0 to top off phenol and then add 0.5 M EDTA, pH 8.0, to a final concentration of 1 mM to prepare the TE saturated phenol (phenol–TE).
    Store at 4 °C in brown glass bottles (*see* **Note** 7).

3.2 Extraction of ATP from Cells and Blood

ATP extraction procedure from blood and cells (Fig. 1).

1. Add 200 μL of blood or cell suspension to 1 mL of premixed and cooled extraction medium of phenol–TE/chloroform/deionized water (6:2:2) reagent at 4 °C in 2 mL microtubes.
2. Mix well by shaking for 20 s and then centrifuge at 10,000 × *g* for 5 min at 4 °C.
3. Transfer the upper aqueous phase (50 μL) into a new microtubes.
4. Dilute the extract stepwise to 10,000-fold for blood and 1–10-fold for culture cells with deionized water.
5. Transfer 10 μL of the diluted extract to assay tube or microplate. Then add 90 μL of firefly luciferin–luciferase reagent and mix or 10 μL of the extract is used for HPLC analysis.
6. Measure the light produced by luminometer. The ATP levels in the cells are normalized by the number of cells or the total hemoglobin in the blood (*see* **Note 8**).

3.3 Extraction of ATP from Animal Tissues

ATP extraction procedure of animal tissues (Fig. 2).

1. Quickly remove 0.1–0.3 g of animal tissue and determine the wet mass in 14 mL of polypropylene tubes (*see* **Note 9**).
2. Homogenate tissue samples with 3 mL of ice-cold phenol–TE using an Ultra-Turrax on ice. Operate three cycles of 30-s homogenization and 30-s cooling on ice (*see* **Note 10**).

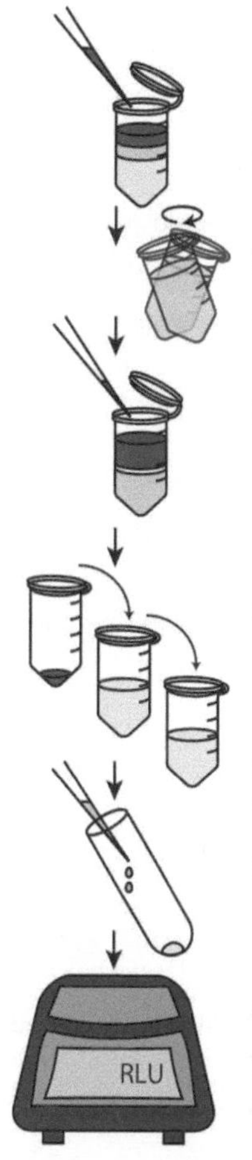

1) Add 200 μL of blood or cell suspension to 1 mL of premixed and cooled extraction media of phenol-TE/chloroform/deionized water (6:2:2) reagent at 4°C in 2 mL microtubes.

2) Mix well by shaking for 20 seconds and then centrifuge at 10,000 × *g* for 5 min at 4 °C.

3) Transfer the upper aqueous phase (50 μL) into a new microtubes.

4) Dilute the extract stepwise to 10,000-fold with deionized water.

5) Transfer 10 μL of the diluted extract to assay tube or microplate. Then add 90 μL of firefly luciferin-luciferase reagent and mix or 10 μL of the extract is used for HPLC analysis.

6) Measure the light produced by luminometer. The ATP levels in the blood or cells are normalized by the number of cells or the total hemoglobin in the blood (*see* **Note 8**).

Fig. 1 Flow scheme illustrating the procedure for ATP extraction from blood and cells

3. Transfer 1.0 mL of the homogenate into 2.0 mL microtube. Then add 200 μL of chloroform and 150 μL of deionized water.
4. Mix well for 20 s and then centrifuge at 10,000 × *g* for 5 min at 4 °C.
5. Collect the upper aqueous phase (50 μL) into a new 2 mL microtube.
6. Dilute the extract stepwise to 100–1,000-fold with deionized water.
7. Transfer 10 μL of the diluted extract to the assay tube or microplate. Then add 90 μL of firefly luciferin–luciferase reagent and mix.
8. Measure the light produced by luminometer. ATP levels in tissues are normalized by the tissue wet weight (*see* **Note 11**).

3.4 Extraction of ATP with TCA

ATP extraction from animal tissues with TCA (*see* **Note 12**).

1. Freshly prepared tissues are immediately homogenized with 3.0 mL of ice-cold homogenization buffer by Ultra-Turrax® using three cycles of 30-s homogenization and 30-s cooling.

1) Quickly remove 0.1 - 0.3 g of animal tissue and determine the wet mass in 14 mL of polypropylene tubes (*see* **Note 8**).

2) Homogenate tissue samples with 3 mL of ice-cold phenol-TE using an Ultra-Turrax on ice. Operate three cycles of 30-seconds homogenization and 30-seconds cooling on ice (*see* **Note 9**).

3) Transfer 1.0 mL of the homogenate into 2.0 mL microtube. Then add 200 μL of chloroform and 150 μL of deionized water.

4) Mix well for 20 seconds and then centrifuge at 10,000 × *g* for 5 min at 4 °C.

5) Collect the upper aqueous phase (50 μL) into a new 2 mL microtube.

6) Dilute the extract stepwise to 100- to 1,000-fold with deionized water.

7) Transfer 10 μL of the diluted extract to the assay tube or microplate. Then add 90 μL of firefly luciferin-luciferase reagent and mix.

8) Measure the light produced by luminometer. ATP levels in tissues are normalized by the tissue wet weight (*see* **Note 10**).

Fig. 2 Flow scheme illustrating the procedure for ATP extraction from mammalian tissues

2. Centrifuge the homogenates at 1,000 × *g* for 10 min at 4 °C. Transfer the supernatant (300 μL) to a new tube.
3. Quickly add 300 μL of ice-cold 10 % TCA followed by shaking for 20 s.
4. Transferred into a 2.0-mL microtube and centrifuged at 10,000 × *g* for 10 min at 4 °C.

5. Added 400 μL of the supernatant to 200 μL of neutralization buffer. The aliquot from the supernatant is diluted 30-fold with deionized water.
6. The diluted extract (10 μL) is used for luciferin–luciferase assay and HPLC analysis.

ATP extraction from blood with TCA (*see* **Note 12**).

1. Blood samples (200 μL) are immediately transferred to 2.0-mL microtubes containing 200 μL of ice-cold 10 % TCA and homogenized or mixed well by vortex mixer.
2. Added to 200 μL of neutralization buffer and centrifuged at 10,000 × *g* for 5 min at 4 °C.
3. The aliquot from the supernatant is diluted 10,000-fold with deionized water, and 10 μL of this diluted extract is used for the analysis.

3.5 Extraction of ATP with EG

ATP extraction from animal tissues with EG (*see* **Note 13**).

1. Freshly prepared tissues are immediately homogenized with 3.0 mL of ice-cold homogenization buffer by Ultra-Turrax® three cycles of 30 s homogenization and 30 s cooling.
2. The homogenate is centrifuged at 1,000 × *g* for 10 min at 4 °C. The supernatant (300 μL) is adjusted to 1.0 mL by the addition of ice-cold homogenization buffer.
3. Quickly added 1.0 mL of ice-cold ATP extraction reagent containing EG provided by the kit (*see* **Note 1**) to the mixture and shaken for 20 s.
4. The mixture is incubated for 30 min at room temperature and 10 μL of this diluted extract is used for analysis.

ATP extraction from blood with EG (*see* **Note 14**).

1. Blood samples are diluted 10,000-fold with dilution buffer provided in the kit.
2. Added an equal volume of ice-cold ATP extraction reagent, and then shaken for 20 s.
3. The extract (10 μL) is used for analysis.

3.6 Luciferin–Luciferase Assay

1. Dilute the ATP stock solution in deionized water to get the optimal detection range between 10^{-7} M and 10^{-12} M before firefly luciferase assay (*see* **Note 15**).
2. Immediately before analysis, add 20 μL of luciferase solution (1 mg/mL) to 6 mL of luciferin–luciferase reaction buffer containing 1 mg of D-luciferin potassium salt to prepare the luciferin–luciferase assay reagent.
3. ATP extract (10 μL) is injected into 90 μL of the luciferin–luciferase assay reagent.

4. Analyzed immediately based on the maximal light intensity of the resulting bioluminescence in a luminometer (TD-20/20).
5. Prepare a calibration curve with ATP standard dilutions and determine ATP levels in each extract from the standard regression curve.
6. The results are normalized against tissue wet weight or protein concentration (*see* **Note 16**).

3.7 HPLC Separation

1. Inject ATP extract (10 μL) to a TSK-GEL Amide-80 reverse-phase chromatography column on a LaChrom D-7000 HPLC System.
2. Separate nucleotides using an isocratic mobile phase at 1 mL/min flow rate at 25 °C. Monitor the eluted nucleotides optically at absorbance of 260 nm.
3. The nucleotide concentration in the eluate is calculated using calibration curves for each standard nucleotide peak area and expressed as μmol/g wet tissue (Fig. 3) (*see* **Note 17**).

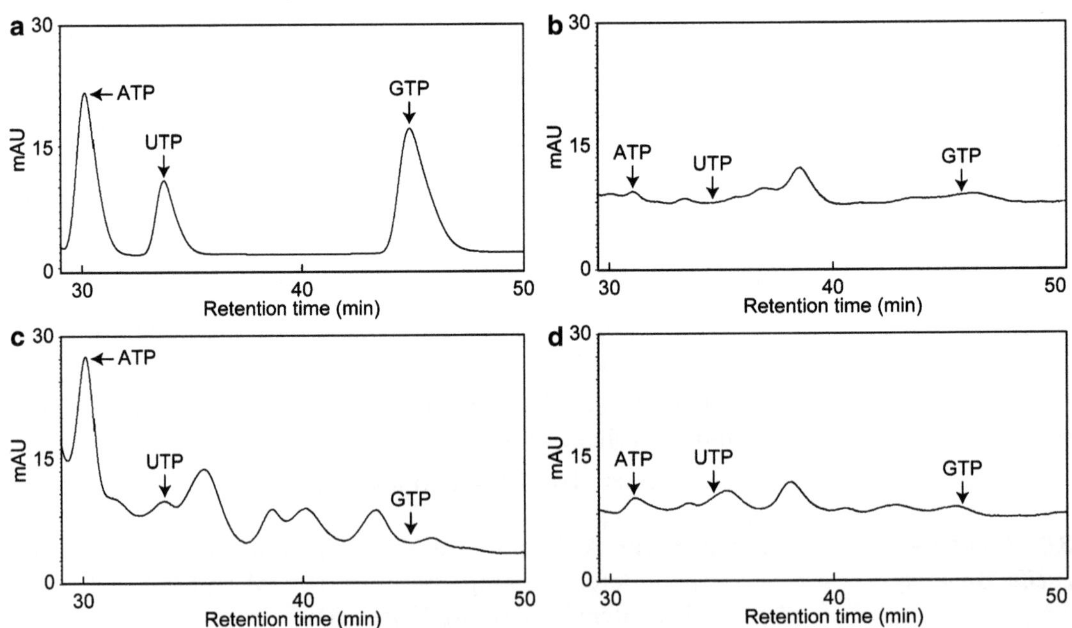

Fig. 3 Separation of nucleotides by HPLC in mouse liver extracts. Spectral view of HPLC chromatogram obtained from a standard mixture of ATP (31.8 pmol), UTP (24.1 pmol), and GTP (47.5 pmol) (**a**). HPLC profiles of nucleotides in the tissues extracted with TCA (**b**), phenol–TE (**c**), and EG (**d**) [14]. Abscissa: retention time, ordinate: absorbance at 260 nm (mAU)

4 Notes

1. ATP extraction medium containing EG from Tissue or Blood ATP measurement Kit®, Toyo Ink, Tokyo.
2. The stock solution is stable for at least 4 weeks when stored at −20 °C.
3. Luciferin and luciferase could be obtained separately or purchased as Firefly Bioluminescence Assay Kits (Enliten® ATP Assay System from Promega, Madison, WI, or AMERIC-ATP kit® and Toyo Ink, Tissue or Blood ATP measurement kit®, Wako Pure Chemical Industries, Osaka, Japan).
4. Phenol is volatile and can cause severe skin burn and damage to clothing. Gloves, safety glasses, and a lab coat should be worn whenever working with phenol, and all manipulations should be carried out in a fume hood.
5. Phenol will be lost during the preparation of the TE buffered phenol. Start with at least 2.5× the final volume of phenol needed in each procedure.
6. If pH is not 8.0 or >7.8, repeat **steps** 7 and **8** until the critical pH is achieved.
7. The buffer saturated phenol may be stored for a period of up to 2 months at 4 °C. TE–saturated phenol is also obtained commercially.
8. The ATP levels in the blood or cells are normalized by protein concentrations of the cell homogenate prepared with TE buffer or appropriate buffer.
9. Alternatively, quick-freeze animal tissues in liquid nitrogen after measurement of tissue weight then grind to a powder.
10. Tissue homogenates in phenol–TE can be stored in dark location at −20 °C for a period up to 6 months without degradation of ATP.
11. Alternatively, ATP levels are normalized by protein concentrations of the tissue homogenate with TE buffer or appropriate buffer.
12. ATP extraction from animal tissues using TCA as a related experiment can be performed according to the protocol supplied by the manufacturer (Enliten® ATP Assay System, Promega, Madison, WI).
13. ATP extraction from animal tissues using EG as a related experiment can be performed according to the protocol supplied by the manufacturer (Tissue ATP measurement Kit®, Toyo Ink, Tokyo).

Table 1
Comparison of ATP level analyzed by firefly bioluminescence assay in mouse tissues and blood extracted with phenol–TE, TCA, and EG

	Extraction method		
Tissues and cells	Phenol–TE (μmol/g wet tissue) ± SD	TCA (μmol/g wet tissue) ± SD	EG (nmol/g wet tissue) ± SD
Brain	0.71 ± 0.06[a]	0.04 ± 0.01	0.29 ± 0.04
Heart	1.43 ± 0.11	0.05 ± 0.01	2.60 ± 1.45
Liver	3.69 ± 0.56	0.01 ± 0.00	1.28 ± 0.23
Spleen	2.55 ± 0.47	0.03 ± 0.00	2.69 ± 1.27
Muscle	10.9 ± 1.90	0.07 ± 0.03	6.41 ± 2.16
Blood	0.76 ± 0.03	0.53 ± 0.01	0.74 ± 0.03

[a]Data are mean ± SD of five experiments

14. ATP extraction from blood using EG as a related experiment is performed according to the protocol supplied by the manufacturer (the Blood ATP measurement Kit®, Toyo Ink, Tokyo).
15. The diluted ATP standard is stable for 8 h when stored on ice.
16. ATP extraction efficiency of several typical extraction reagents, such as phenol–TE, TCA, and EG, by luciferin–luciferase assay in various mouse tissues is shown in Table 1 [14]. The data indicate that phenol–TE extraction is the best extraction method of those tested, producing 17.8-fold and 156-fold higher ATP levels than TCA in the brain and muscle, respectively, and 550–1,000-fold higher than EG in the heart and other mouse tissues with a relatively high protein concentration. However, there are no significant variability in ATP level in cultured cells and bacteria with a relatively low protein concentration among the three extraction methods.
17. The HPLC profiles indicate that ATP is well separated from other nucleotides. Table 2 shows the extraction efficiency of ATP and other nucleotides from mice liver extracted with phenol-TE, TCA, and EG [14]. Although these extraction reagents have different nucleotides extraction efficiencies, the phenol–TE method shows the best overall extraction efficiency for ATP, UTP, and GTP from the mouse tissues tested.

Acknowledgments

This work was supported by a Grant-in-Aid for Exploratory Research (No. 21790992) from J.S.P.S.

Table 2
Comparison of nucleotide levels analyzed by HPLC of mouse tissues extracted with phenol–TE, TCA, and EG

	Extraction method		
Tissues and nucleotides	Phenol–TE (μmol/g wet tissue) ± SD	TCA (μmol/g wet tissue) ± SD	EG (nmol/g wet tissue) ± SD
Brain			
ATP	0.58 ± 0.22	0.04 ± 0.03	n.d.
UTP	0.24 ± 0.35	0.08 ± 0.04	1.03 ± 0.07
GTP	0.70 ± 0.19	0.24 ± 0.08	0.98 ± 0.43
Heart			
ATP	2.07 ± 0.71	0.04 ± 0.04	1.84 ± 1.04
UTP	n.d.	n.d.	n.d.
GTP	2.26 ± 0.57	0.56 ± 0.14	2.96 ± 0.75
Liver			
ATP	2.41 ± 0.76	0.02 ± 0.02	2.02 ± 0.61
UTP	n.d.	n.d.	n.d.
GTP	0.35 ± 0.24	0.03 ± 0.04	n.d.
Muscle			
ATP	10.5 ± 0.66	0.08 ± 0.02	1.02 ± 0.97
UTP	0.92 ± 0.50	0.27 ± 0.06	n.d.
GTP	n.d.	n.d.	11.2 ± 1.17

n.d. not detectable
Data are mean ± SD of three experiments

References

1. Nilsson L, Kogure K, Busto R (1975) Effects of hypothermia and hyperthermia on brain energy metabolism. Acta Anaesthesiol Scand 19:199–205
2. Zhu H, Zennadi R, Xu BX et al (2011) Impaired adenosine-5′-triphosphate release from red blood cells promotes their adhesion to endothelial cells : a mechanism of hypoxemia after transfusion. Crit Care Med 39:2478–2486
3. Glembotski CC, Chapman AG, Atkinson DE et al (1981) Adenylate energy charge in *Escherichia coli* CR341T28 and properties of heat-sensitive adenylate kinase. J Bacteriol 145: 1374–1385
4. Calderwood SK, Bump EA, Stevenson MA et al (1985) Investigation of adenylate energy charge, phosphorylation potential, and ATP concentration in cells stresses with starvation and heat. J Cell Physiol 124:261–268
5. Fedorow CA, Churchill TA, Kneteman NM (1998) Effects of hypothermic hypoxia on anaerobic energy metabolism in isolated anuran livers. J Comp Physiol B 168:555–561
6. Kohane MJ, Watt WB (1999) Flight-muscle adenylate pool responses to flight demands and thermal constraints in individual Colias eurytheme (Lepidoptera, pieridae). J Exp Biol 202: 3145–3154
7. Stanley PE, Williams SG (1969) Use of the liquid scintillation spectrometer for determining adenosine triphosphate by the luciferase enzyme. Anal Biochem 29:381–392

8. Lundin A, Thore A (1975) Comparison of methods for extraction of bacterial adenine nucleotides determined by firefly assay. Appl Microbiol 30:713–721
9. Thore A, Anséhn S, Lundin A et al (1975) Detection of bacteriuria by luciferase assay of adenosine triphosphate. J Clin Microbiol 1: 1–8
10. Napolitano MJ, Shain DJ (2005) Quantitation adenylate nucleotides in diverse organisms. J Biochem Biophys Methods 63:69–77
11. St.John JB (1970) Determination of ATP in Chlorella with the luciferin-luciferase enzyme system. Anal Biochem 37:409–416
12. Williams C, Forrester T (1976) Loss of ATP in micromolar amounts after perchloric acid treatment. Pflügers Arch 366:281–283
13. Wiener S, Wiener R, Urivetzky M et al (1974) Coprecipitation of ATP with potassium perchlorate : the effect of the firefly enzyme assay of ATP in tissue and blood. Anal Biochem 59:489–500
14. Chida J, Yamane K, Takei T et al (2012) An efficient extraction method for quantitation of adenosine triphosphate in mammalian tissues and cells. Anal Chim Acta 727:8–12
15. Sambrook J, Eritch EF, Maniatis T (1989) Molecular cloning, a laboratory manual. Cold Spring Harbor Laboratory, New York

Chapter 3

Neuronal Network Imaging in Acute Slices Using Ca^{2+} Sensitive Bioluminescent Reporter

Ludovic Tricoire and Bertrand Lambolez

Abstract

Genetically encoded indicators are valuable tools to study intracellular signaling cascades in real time using fluorescent or bioluminescent imaging techniques. Imaging of Ca^{2+} indicators is widely used to record transient intracellular Ca^{2+} increases associated with bioelectrical activity. The natural bioluminescent Ca^{2+} sensor aequorin has been historically the first Ca^{2+} indicator used to address biological questions. Aequorin imaging offers several advantages over fluorescent reporters: it is virtually devoid of background signal; it does not require light excitation and interferes little with intracellular processes. Genetically encoded sensors such as aequorin are commonly used in dissociated cultured cells; however it becomes more challenging to express them in differentiated intact specimen such as brain tissue. Here we describe a method to express a GFP-aequorin (GA) fusion protein in pyramidal cells of neocortical acute slices using recombinant Sindbis virus. This technique allows expressing GA in several hundreds of neurons on the same slice and to perform the bioluminescence recording of Ca^{2+} transients in single neurons or multiple neurons simultaneously.

Key words Aequorin, Calcium imaging, Neocortex, Sindbis, Bioluminescence

1 Introduction

1.1 Bioluminescent Genetically Encoded Ca^{2+} Indicator

Aequorin, isolated from jellyfish Aequorea species, is a bioluminescent complex that emits blue light upon Ca^{2+} binding [1]. Aequorin was first used as a Ca^{2+} indicator to evidence the role of intracellular Ca^{2+} in excitation–contraction coupling [2], neuronal signaling [3, 4], and meosis [5]. Since then, aequorin has been extensively used as a reporter of Ca^{2+} physiology in various cell types and subcellular compartments following intracellular injection or expression by gene transfer (reviewed in ref. 6).

Aequorin belongs to the family of photoproteins, i.e., it is a stable luciferase intermediate formed from the reaction of the protein apoaequorin (luciferase) and the prosthetic group coelenterazine (luciferin), and contains three EF-hand Ca^{2+}-binding sites [7–9].

Christian E. Badr (ed.), *Bioluminescent Imaging: Methods and Protocols*, Methods in Molecular Biology, vol. 1098,
DOI 10.1007/978-1-62703-718-1_3, © Springer Science+Business Media New York 2014

The formation of the aequorin complex is a slow process, whereas the bioluminescence reaction occurs as a rapid flash whose intensity increases with Ca^{2+} concentration and proceeds to completion in the continuous presence of Ca^{2+} [7, 10–12]. The rapid kinetics of Ca^{2+} binding to and unbinding from aequorin makes it a suitable indicator of rapid Ca^{2+} transients [10]. Aequorin Ca^{2+} sensitivity span concentrations from 0.1 to 1 mM [12, 13]. In addition, aequorin mutants and semisynthetic aequorins (i.e., incorporating synthetic coelenterazine analogue) endowed with different kinetics, Ca^{2+} sensitivity or spectral properties [12, 14–17] expand the range of bioluminescent Ca^{2+} sensors available to diverse applications [18–20].

A GFP-aequorin (GA) fusion protein has been described that allows fluorescence labeling of expressing cells and bioluminescence Ca^{2+} imaging of single cultured cells, of tissue slices and in whole vertebrate and invertebrate animals [21–25]. The fact that no illumination is required has allowed the use of GA to report neuronal activity in freely behaving zebrafishes [23]. No conspicuous morphological or functional alteration has been observed upon acute high-level expression of GA or in transgenic animals stably expressing GA [22–24]. Its bioluminescence *in cellulo* is superior to that of aequorin alone, presumably because the latter is rapidly degraded in the cytoplasm [26]. Bioluminescence resonance energy transfer between the photoprotein and GFP moieties of both fusion proteins shifts light emission to the green [27, 28] as observed between native fluorescent proteins and photoproteins in light emitting cells of *Aequorea victoria* [29]. Further shift in emission spectra has been observed when aequorin was fused to yellow or red fluorescent proteins enabling light detection though intact skin on living animals [30, 31].

Bioluminescence imaging with photoproteins is endowed with a high signal-to-noise ratio. Indeed, photoproteins exhibit negligible Ca^{2+}-independent luminescence and their response intensity varies by several orders of magnitude depending on Ca^{2+} concentration [13, 32, 33]. GFP-photoproteins behave as low affinity, supralinear indicators of Ca^{2+} transients associated with action potentials in mammalian neurons [21]. A detection threshold of five action potentials has been reported in cortical neurons upon bioluminescence imaging in brain slices, corresponding to Ca^{2+} concentration transients locally reaching the micromolar range [21]. This detection threshold is comparable to that reported for the GCaMP genetically encoded fluorescent Ca^{2+} sensor [34]. The substrate coelenterazine is generally loaded once for reconstitution of active photoprotein prior to imaging experiments. The absence of coelenterazine during the course of the experiment eventually leads to exhaustion of the active photoprotein. However, due to the relatively small percentage of aequorin consumed, long duration continuous recordings of several hours have been reported [23].

1.2 Sindbis-Based Gene Transfer into Neurons

Sindbis virus belongs to the family of alphaviruses, in that its genome consist of a single-stranded, plus-strand RNA genome with a 5′ cap and a 3′ poly(A) tail. Upon entry into the cell, the viral RNA is directly translated by the cell host machinery and a virally encoded RNA-dependent RNA polymerase amplifies the genomic RNA, and generates transcripts that are under the control of a subgenomic promoter within the genomic RNA.

SIN virus has been successfully transformed into expression vectors, in which the transgene in the "vector RNA" is located downstream the subgenomic promoter and replaces the viral structural protein genes [35]. Shortly after cell transduction, the subgenomic RNA becomes the most abundant message in the transduced cells and promptly recruits most of the host's translational machinery for its own use, resulting in high levels of the desired protein molecules in the cytoplasm few hours after transduction. Such expression SIN vectors are called pseudovirions as they do not encode for structural proteins such as capsid and have little or no plaque forming unit capability. For more details about the biology of Sindbis vectors *see* ref. [36].

Compared to other viral vectors, advantages of SIN vectors are available plasmids for cloning the transgene and synthesis of the recombinant vector RNA, large insert size (6.5 kb), easy and fast generation of recombinant viral particles, high viral titers obtained, rapid onset, and high-level transgene expression (within hours) [37]. When applied to neuronal tissue, it possesses the additional advantage of efficiently and preferentially transducing neurons rather than nonneuronal cells [36]. In recent years, Sindbis viruses have been used to express various membrane receptors, anchoring molecules and kinases in cultured neurons and cultured slices, and also in vivo [36]. Furthermore, several genes can be expressed separately using a multiple subgenomic promoter [38].

Aside from their advantages, wild-type SIN pseudovirion vectors [35] retain the disadvantage of inhibiting host cell protein synthesis, which eventually causes cell death. Therefore, SIN expression systems are often used for transient expression experiments. While neurotoxic effects appear relatively fast (within 1–2 days), some mutation in the nonstructural protein 2 or the use of optimized helper RNA have been found to delay the onset of cellular toxicity by several days [39, 40]. In addition of SIN pseudovirions, other expression vectors based on the semliki forest virus (SFV), another alphavirus, have been successfully assembled [36] which allow targeting either a specific neuronal cell types or nonneuronal cells [36]. This can be achieved for example by engineering coat protein [41] or using microRNA [42].

In this chapter we describe a protocol to prepare acute slice for SIN viral infection and to perform subsequent imaging of the GA light response upon neuronal activity evoked by electrical stimulation. It must be noted that we routinely use SIN pseudovirions generated using the vector plasmid pSinRep5 [35] to express several types of genetically encoded indicators such as GA into neocortical pyramidal cell or thalamic neurons of acute slices [21, 43–45].

2 Materials

Experimental animals: rats (Wistar) or mice (C57/Bl6). All animal experiments are to be performed in accordance with the guidelines on the use of animals by the relevant authorities.

CAUTION: The low level of pathogenicity of SIN virus in humans has allowed it to be classified as a Biosafety Level-2 (BL-2) agent by the NIH Recombinant DNA Advisory Committee. Recommended precautions include standard microbiological practices, laboratory coats, inactivation of all infectious waste, limited access to working areas, protective gloves, posted biohazard signs, and class I or II biological safety cabinets used for mechanical and manipulative procedures that cause splashes or aerosol. All personnel working with the SIN Expression System should be properly trained to work with BL-2 level organisms. SIN virus can be inactivated by organic solvents, bleach, or autoclaving. SIN vectors expressing a highly toxic protein should be treated as a special risk.

2.1 Preparation of Acute Slices

1. 10× ACSF (in mM): 1,260 NaCl, 25 KCl, 12.5 NaH_2PO_4, 260 $NaHCO_3$. Weigh 146.1 g NaCl, 3.73 g KCl, 3 g anhydrous NaH_2PO_4, 43.7 g $NaHCO_3$, and add to a 2 l-cylinder containing 1 l distilled water and make it up to 2 l. Mix well and store at 4 °C. This solution can be kept for 2 weeks.
2. 1× ACSF (in mM): 126 NaCl, 2.5 KCl, 1.25 NaH_2PO_4, 2 $CaCl_2$, 1 $MgCl_2$, 26 $NaHCO_3$, 20 D-glucose, and 5 Na pyruvate. Add in this order, 750 ml volvic water, D-glucose, 100 ml 10× ACSF, 2 ml $CaCl_2$ 1 M, 1 ml $MgCl_2$ 1 M and 0.55 g Na pyruvate. Adjust to 1 l and filter this solution on 0.22 μm membrane. This solution is prepared every day and saturated with carbogen (95 % O_2, 5 % CO_2) for 20 min before use.
3. The cutting solution consists in ACSF supplemented with 1 mM kynurenic acid. It is also bubbled with carbogen. To help dissolving kynurenic acid, sonicate for 5–10 min the ACSF solution containing kynurenic acid using an ultrasound bath.
4. Dissection tools for removing the brain. This includes Scissors for initial dissection, long thin scissors for removing skin, short scissors for cutting skull, short blunt forceps for removing skull, rounded spatula for removing brain, scalpel and blade for dividing brain, rounded spatula for transferring brain halves.
5. Small parts: Adhesive cyanoacrylate (e.g., Loctite 406), Stainless steel blade or razor blade, plastic Petri dishes (diameter 35 mm), Pasteur pipette for transferring slices.

6. Anesthetics ketamine (100 mg/kg body weight) and xylasine (10 mg/kg body weight). In a 5 ml tube, add 125 μl xylasine (e.g., Rampun 0.2 %), 250 μl ketamine (e.g., Imalgene 1000), and 2.125 ml NaCl 0.9 %. Store this solution at 4 °C.

2.2 Slice Culture

1. In a 500 ml cylinder, add 250 ml Minimal essential medium (MEM), 250 ml 1× Hanks balanced salt solution (HBSS), 5 ml Penicillin–streptomycin (P: 10,000 U/ml, S: 10,000 mg/ml), 2.75 g D-glucose (6.5 g/l final) then filter-sterilized before aliquoting in 50 ml tube (*see* **Note 1**).
2. Millicell culture plate inserts (Millipore); these inserts have low walls, which help to the handling of the slices and allow putting the insert in a 35 mm Petri dish.
3. Recombinant Sindbis pseudovirus. A detailed protocol for the production of Sindbis-based vectors can be found elsewhere [46]. We routinely use SIN vectors assembled using pSinRep5 vector plasmids and the DH26S helper plasmid [35]. Briefly, the coding sequence of the gene of interest is inserted in pSinRep5 downstream of the subgenomic promoter. The resulting plasmids and pDH26S are then linearized and in vitro transcribed into capped RNA using SP6 polymerase. Both RNAs are electroporated into BHK-21 cells and supernatant containing SIN pseudovirion is collected 24 h later.
4. Native coelenterazine free-base, stock solution at 1.25 mM in ethanol. Stored at −80 °C. Caution: manipulate coelenterazine in the dark.

2.3 Equipment

1. Laminar flow cabinet and a 5 % CO_2 incubator maintained at 35 °C.
2. High-quality vibrating slicer (VT1000S or VT1200; Leica Microsystems).
3. Incubation chamber: a submerged chamber optimized to ensure sufficient oxygenation of the tissue during the recovery period. Usually, it consists of a small plastic cylinder ending in a nylon net.
4. Grid of nylon threads glued to a U-shaped platinum frame to maintain the slice in the bottom of the recording chamber. The platinum U-frame should be centered to avoid any mechanical interaction between its walls and the recording pipettes.
5. An imaging setup that can contain an upright microscope with water immersion objective, a recording chamber and a perfusion system. Example of such equipment can be found elsewhere [47, 48]. We are using an intensified CCD video camera (ICCD225; 768×576 pixels; Photek, St Leonards on Sea, UK) mounted on the C-mount port of an upright BX51WI microscope (Olympus) and controlled by the data acquisition

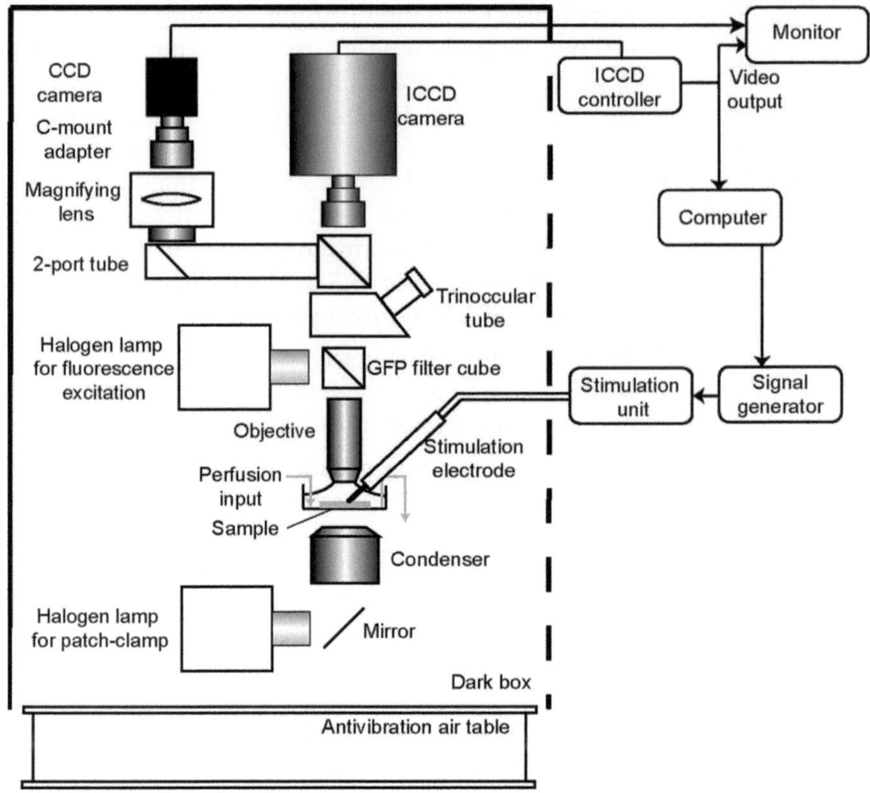

Fig. 1 Schematic representation of a setup for bioluminescence recordings. Prior to imaging, samples are visualized by GFP epifluorescence using GFP filter set. Filters reside in a filter wheel (not shown), thus they can be removed from the light path during bioluminescence acquisition. Fluorescence image can be acquired either with the low gain mode of the ICCD camera or using a regular CCD camera connected to a 2-tube port

software IFS32 (Photek). Light acquisition is performed through water immersion objectives 10× (N.A. = 0.3) and 60× (N.A. = 0.9). The imaging setup (*see* Fig. 1) is housed in a dark box to avoid light noise. Slices are set in a recording chamber that is continuously perfused with standard artificial cerebrospinal fluid (ACSF). GFP fluorescence is visualized using a mercury lamp with a standard GFP filter set (e.g., from Chroma technology, cat no 41018, or Semrock, cat no GFP-30LP-B-000). Filters are removed from light path during bioluminescence recordings.

6. For electrical stimulation, a concentric bipolar electrode (SNEX-100, Rhode biomedical, tip end 100 μm wide) is set up on a manually driven micromanipulator (Mini 25 3Axes, Luigs and Neuman, Germany). The electrode is connected to a constant current stimulation unit like the DS2A from Digitimer (UK). Pulse amplitude and duration are set on the stimulation unit. The stimulus wave form is provided by a

TTL signal either by the software pClamp through a digidata A/D converter or with a signal generator (Accupulse, World Precision Instrument).

3 Methods

3.1 Brain Dissection and Acute Slice/Slice Culture Preparation

1. Prepare the slicing setup by putting 250 ml of cutting solution in the freezer until ice appears (*see* **Note 2**). Transfer this solution to a polystyrene box containing ice and saturate the liquid/ice mixture with carbogen for 20 min. During that period, install dissection instruments on the bench, set up the razor blade on the blade holder of the slicing microtome and put some ice around the slicing chamber in which the slice will be cut. Some water can be added to the ice to facilitate the cooling of the tray and to keep it cold along the slicing procedure. Prepare an inverted suction pipette to water the brain with a Pasteur pipette broken at the narrow tip and mounted on a rubber bulb.
2. Anesthetize the animal with a ketamine xylazine mixture. We usually use 13–15-day-old rats and mice for acute cortical slice to be infected with recombinant Sindbis virus. Sacrifice the animal by decapitation with either a scissor or a guillotine at the level of the medulla. Submerge the head in a Petri dish containing icy cutting solution saturated with carbogen. With a scalpel blade, cut the scalp bilaterally in a caudal direction from above the eyes toward the posterior of the skull. Push the scalp aside and using small, thin scissors, cut the skull bilaterally in a rostral direction from the vertebral foramen toward the frontal lobes. Make four lateral cuts from the midline, two anterior and two posterior. Be sure these final cuts are sufficiently anterior and posterior so as not to damage the brain. Using blunt forceps, open the skull from the midline in a lateral direction. During this step, check that there are no meninges still attached to the skull as it may damage the cortex. While cutting the skull, keep the tip of the scissors as close to the skull as possible with upward motion. The extraction of the brain should be accomplished within ~1–2 min after decapitation.
3. For parasagittal section, using a scalpel blade, make two lateral cuts between the olfactory bulb and the frontal cortex, and between the cerebellum and the occipital cortex, then isolate the two hemispheres and extract the two half brains from the skull with a spatula. Slide the base of the tissue block from the spatula onto a thin film of cyanocrylate glue on the cooled slicing stage. Due to the curvature of the brain surface, we use a homemade slicing stage with a 10° slope. The rostro caudal axis is parallel to the blade edge and the dorsal part of the brain is oriented downward (*see* Fig. 2).

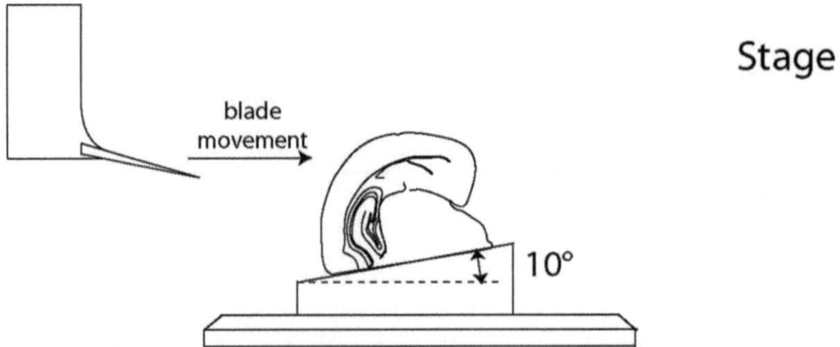

Fig. 2 10° slicing stage

4. Transfer the slicing stage in the slicing chamber, screw it and submerge the stage with the block of tissue by pouring saturated cold cutting solution (*see* **Note 3**). Cut slices at 300 μm (250 μm for mice). The vibration rate of the blade is usually fast (~60–70 Hz or graduation 6–7 on the VT1000S) and the speed is nearly minimal (0.1–0.2 mm/s or graduation 1–2 on the VT1000S). The amplitude of the horizontal blade movement is 0.8 mm. During the slicing procedure, the solution must be continuously bubbled with carbogen with a minimal flow rate to avoid turbulence that could damage the slices.
5. Once the blade has been through the cortex and the hippocampus, make a horizontal cut with a scalpel blade or a folded thin needle set on a syringe to release the slice and restart sectioning. This allows saving time by not continuing cutting tissue containing more ventral structures and restricts the slice to the regions of interest. Transfer the slices which should contain cortex, hippocampus and a piece of dorsal striatum into the incubation chamber containing cutting solution and let them recover for 30 min at room temperature (*see* **Note 4**).
6. During the recovery period, add 1.2 ml of MEM-HBSS medium in a 35 mm Petri dish and put one millicell insert. Make sure no bubbles remain under the insert. Move the Petri dish to a 5 % CO_2 incubator until slices transfer.
7. Transfer the slices using the inverted Pasteur pipette. Remove the maximum of supernatant to leave a thin layer of medium above the slice (*see* **Note 5**).
8. Slices are transduced with Sindbis pseudovirions by adding 5 μl of viral solution onto the slice and incubating at 35 °C in a standard tissue culture incubator with an humidified 5 % CO_2 atmosphere (*see* **Note 6**).
9. 30 min after virus application, coelenterazine (final concentration: 10 μM) is added to the culture medium and slices are put back in the incubator and left overnight until the recording.

3.2 Bioluminescence Imaging

1. The next day, slices are transferred into an incubation chamber containing carbogen-saturated 1× ACSF, incubated for at least 1 h at room temperature in the dark. The incubation step in ACSF is necessary to equilibrate the slice with the recording solution and to avoid drift along the *z*-axis during recording (*see* **Note 7**).
2. Transfer one slice into the recording chamber and perfuse at 1–2 ml/min with saturated ACSF. Before the transfer, stop the perfusion by interruption suction and inflow, put delicately the slice at the bottom of the recording chamber and place a grid onto the slice and resume the perfusion of the chamber. The grid is necessary to avoid lateral movement of the slice during the recording due to the perfusion of the chamber especially if the chamber volume is small. Recordings are typically performed 15–24 h after viral transduction. Recording can be performed either at room temperature or 32–34 °C.
3. Using a low magnification objective (e.g., 10× immersion objective) and by moving the stage or the recording chamber, scan the slice to locate a region with the desired expression pattern (*see* **Note 8**).
4. Using the micromanipulator, position the tip of the stimulation electrode on the top of the slice. The electrode should not penetrate the slice as it may damage tissue.
5. Change the objective if necessary and using the low gain acquisition mode of the camera, take fluorescence image of the imaging field. This will be used at the end of the experiment to check if there has been some lateral or vertical drift of the slice during the recording.
6. Close the dark box and turn off any unnecessary source of light in the room. Remove the GFP filter cube from the light path and switch on the high gain mode of the camera and start the recording.
7. At the end of the recording it can be necessary to evoke large intracellular Ca^{2+} increase for calibration between slices. This can be achieved with calcium ionophore such as ionomycin, but our experience indicates that its penetration into the slice is poor. We suggest applying glutamate 100–500 μM.
8. Save the file and switch back to the low gain mode and take a new fluorescence picture of the imaging field.
9. If histological analysis is necessary, transfer the slice into a 24-well plate containing 4 % paraformaldehyde in PBS and incubate overnight at 4 °C. The day after, replace the fixative with PBS.
10. For analysis, some simple measurements can be performed using the acquisition program IFS32. However if more

complex calculations are necessary, we developed a program to convert property files generated by IFS32 into TIFF movie that can be then opened in any image analysis software.

4 Notes

1. The MEM contains HEPES and Phenol red but no L-glutamine. The HBSS contains calcium, magnesium, phenol red, glucose, and sodium bicarbonate. This solution can be kept at 4 °C for several weeks. Once opened the culture medium contained in the 50 ml tube has to be used within 2 weeks.
2. It is important that the ACSF remains ice cold during all the procedure. Especially, just after the decapitation, the head and the brain must be quickly cooled down. During the slicing, it is useful to have small ice cubes of ACSF in the slicing chamber.
3. When pouring ACSF on the tissue block, make sure no ice cube falls on the brain otherwise it will damage the tissue.
4. This protocol is derived from the slicing procedure used for standard patch-clamp recording techniques. However, given that the slices have to survive overnight to allow transgene expression, the quality of the slice surface and cell survival is very important. For those who are not familiar with brain slice recordings, it is advisable to train before performing viral transduction. Check the slice quality using a regular upright microscope such as those used for electrophysiological experiments. The duration of the whole slicing procedure should be as minimal as possible (typically less than 20 min). The time between decapitation and transfer of the brain is cold ACSF should less than 2 min.
5. Avoid putting more than two slices per millicell insert. Volume of medium under the millicell may have to be adjusted depending on the Petri dish brand as slice may be completely soaked. If it is the case, decrease volume but to no less than 1 ml.
6. Volume of viral solution can be adjusted depending on the area to be transduced. Virus can be injected locally for a more focal expression. Make sure that virus aliquot is at room temperature before adding to the slice.
7. Usually the slices are not fully attached to the insert membrane and thus can be detached easily. We suggest to pour some temperature-equilibrated carbogen-saturated ACSF on the slice using an inverted Pasteur pipette and then to detach the slice using a thin paintbrush. Given that the Sindbis virus does not diffuse deeply into the slice (<150 μm), it is important to keep the infected side of the slice upward.

8. Several reasons can explain lack of fluorescent cells:
    - You are examining the wrong side of the slice (*see* **Note 6**).
    - Viral titer is not high enough. We usually use minimum 10^7 infectious particles per ml as measured by infecting BHK-21 cells.
    - The volume of MEM-HBSS is too high and oxygen diffusion is not good enough.
    - The viral suspension was too cold when added to the slice.
    - The slice was not of sufficient quality and neurons died quickly after slicing. Depending of the brain structure of interest, you may have to change the composition of the cutting ACSF or the culture medium. We had equivalent results with a mix Neurobasal-A+B27 supplement instead of HBSS+MEM. For thalamic neurons we replace the cutting ACSF by a PIPES and sucrose-based solution (*see* ref. 44).

References

1. Shimomura O, Johnson FH, Saiga Y (1962) Extraction, purification and properties of aequorin, a bioluminescent protein from the luminous hydromedusan, Aequorea. J Cell Comp Physiol 59:223–239
2. Ridgway EB, Ashley CC (1967) Calcium transients in single muscle fibers. Biochem Biophys Res Commun 29(2):229–234. doi:0006-291X(67)90592-X [pii]
3. Llinas R, Blinks JR, Nicholson C (1972) Calcium transient in presynaptic terminal of squid giant synapse: detection with aequorin. Science 176(39):1127–1129
4. Stinnakre J, Tauc L, Saito N (1972) [Demonstration with aequorine photoprotein of variations of calcium activity in Aplysia neurons]. J Physiol (Paris) 65:Suppl:308A
5. Moreau M, Guerrier P et al (1978) Hormone-induced release of intracellular Ca2+ triggers meiosis in starfish oocytes. Nature 272(5650): 251–253
6. Pinton P, Rimessi A, Romagnoli A et al (2007) Biosensors for the detection of calcium and pH. Methods Cell Biol 80:297–325. doi:S0091-679X(06)80015-4 [pii] 10.1016/S0091-679X(06)80015-4
7. Shimomura O, Johnson FH (1975) Regeneration of the photoprotein aequorin. Nature 256(5514):236–238
8. Vysotski ES, Lee J (2004) Ca2+-regulated photoproteins: structural insight into the bioluminescence mechanism. Acc Chem Res 37(6):405–415. doi:10.1021/ar0400037
9. Wilson T, Hastings JW (1998) Bioluminescence. Annu Rev Cell Dev Biol 14:197–230. doi:10.1146/annurev.cellbio.14.1.197
10. Hastings JW, Mitchell G, Mattingly PH et al (1969) Response of aequorin bioluminescence to rapid changes in calcium concentration. Nature 222(5198):1047–1050
11. Shimomura O, Johnson FH (1970) Calcium binding, quantum yield, and emitting molecule in aequorin bioluminescence. Nature 227(5265):1356–1357
12. Tricoire L, Tsuzuki K, Courjean O et al (2006) Calcium dependence of aequorin bioluminescence dissected by random mutagenesis. Proc Natl Acad Sci U S A 103(25):9500–9505. doi:0603176103 [pii] 10.1073/pnas.0603176103
13. Allen DG, Blinks JR, Prendergast FG (1977) Aequorin luminescence: relation of light emission to calcium concentration–a calcium-independent component. Science 195(4282):996–998
14. Kendall JM, Sala-Newby G, Ghalaut V et al (1992) Engineering the CA(2+)-activated photoprotein aequorin with reduced affinity for calcium. Biochem Biophys Res Commun 187(2):1091–1097
15. Shimomura O, Musicki B, Kishi Y (1989) Semi-synthetic aequorins with improved sensitivity to Ca2+ ions. Biochem J 261(3): 913–920
16. Stepanyuk GA, Golz S, Markova SV et al (2005) Interchange of aequorin and obelin bioluminescence color is determined by substitution of one active site residue of each

photoprotein. FEBS Lett 579(5):1008–1014. doi:S0014-5793(05)00033-5 [pii] 10.1016/j.febslet.2005.01.004

17. Tsuzuki K, Tricoire L, Courjean O et al (2005) Thermostable mutants of the photoprotein aequorin obtained by in vitro evolution. J Biol Chem 280(40):34324–34331. doi:M505303200 [pii] 10.1074/jbc.M505303200
18. Llinas R, Sugimori M, Silver RB (1992) Microdomains of high calcium concentration in a presynaptic terminal. Science 256(5057):677–679
19. Llinas R, Sugimori M, Silver RB (1995) The concept of calcium concentration microdomains in synaptic transmission. Neuropharmacology 34(11):1443–1451. doi:0028390895001505 [pii]
20. Robert V, De Giorgi F, Massimino ML et al (1998) Direct monitoring of the calcium concentration in the sarcoplasmic and endoplasmic reticulum of skeletal muscle myotubes. J Biol Chem 273(46):30372–30378
21. Drobac E, Tricoire L, Chaffotte AF et al (2010) Calcium imaging in single neurons from brain slices using bioluminescent reporters. J Neurosci Res 88(4):695–711. doi:10.1002/jnr.22249
22. Martin JR, Rogers KL, Chagneau C, Brulet P (2007) In vivo bioluminescence imaging of Ca signalling in the brain of Drosophila. PLoS One 2(3):e275. doi:10.1371/journal.pone.0000275
23. Naumann EA, Kampff AR, Prober DA et al (2010) Monitoring neural activity with bioluminescence during natural behavior. Nat Neurosci 13(4):513–520. doi:nn.2518 [pii] 10.1038/nn.2518
24. Rogers KL, Picaud S, Roncali E et al (2007) Non-invasive in vivo imaging of calcium signaling in mice. PLoS One 2(10):e974. doi:10.1371/journal.pone.0000974
25. Rogers KL, Stinnakre J, Agulhon C et al (2005) Visualization of local Ca2+ dynamics with genetically encoded bioluminescent reporters. Eur J Neurosci 21(3):597–610. doi:EJN3871 [pii] 10.1111/j.1460-9568.2005.03871.x
26. Badminton MN, Sala-Newby GB, Kendall JM, Campbell AK (1995) Differences in stability of recombinant apoaequorin within subcellular compartments. Biochem Biophys Res Commun 217(3):950–957. doi:S0006291X85728628 [pii]
27. Baubet V, Le Mouellic H, Campbell AK et al (2000) Chimeric green fluorescent protein-aequorin as bioluminescent Ca2+ reporters at the single-cell level. Proc Natl Acad Sci U S A 97(13):7260–7265. doi:97/13/7260 [pii]
28. Gorokhovatsky AY, Marchenkov VV, Rudenko NV et al (2004) Fusion of Aequorea victoria GFP and aequorin provides their Ca(2+)-induced interaction that results in red shift of GFP absorption and efficient bioluminescence energy transfer. Biochem Biophys Res Commun 320(3):703–711. doi:10.1016/j.bbrc.2004.06.014 S0006291X04012793 [pii]
29. Morin JG, Hastings JW (1971) Energy transfer in a bioluminescent system. J Cell Physiol 77(3):313–318. doi:10.1002/jcp.1040770305
30. Bakayan A, Vaquero CF, Picazo F, Llopis J (2011) Red fluorescent protein-aequorin fusions as improved bioluminescent Ca2+ reporters in single cells and mice. PLoS One 6(5):e19520. doi:10.1371/journal.pone.0019520 PONE-D-10-04882 [pii]
31. Curie T, Rogers KL, Colasante C, Brulet P (2007) Red-shifted aequorin-based bioluminescent reporters for in vivo imaging of Ca2 signaling. Mol Imaging 6(1):30–42
32. Illarionov BA, Frank LA, Illarionova VA et al (2000) Recombinant obelin: cloning and expression of cDNA purification, and characterization as a calcium indicator. Methods Enzymol 305:223–249
33. Markova SV, Vysotski ES, Blinks JR et al (2002) Obelin from the bioluminescent marine hydroid Obelia geniculata: cloning, expression, and comparison of some properties with those of other Ca2+-regulated photoproteins. Biochemistry 41(7):2227–2236. doi:bi0117910 [pii]
34. Pologruto TA, Yasuda R, Svoboda K (2004) Monitoring neural activity and [Ca2+] with genetically encoded Ca2+ indicators. J Neurosci 24(43):9572–9579. doi:24/43/9572 [pii] 10.1523/JNEUROSCI.2854-04.2004
35. Bredenbeek PJ, Frolov I, Rice CM, Schlesinger S (1993) Sindbis virus expression vectors: packaging of RNA replicons by using defective helper RNAs. J Virol 67(11):6439–6446
36. Ehrengruber MU (2002) Alphaviral gene transfer in neurobiology. Brain Res Bull 59(1):13–22. doi:S0361923002008584 [pii]
37. Washbourne P, McAllister AK (2002) Techniques for gene transfer into neurons. Curr Opin Neurobiol 12(5):566–573. doi:S0959438802003653 [pii]
38. Okada T, Yamada N, Kakegawa W et al (2001) Sindbis viral-mediated expression of Ca2+-permeable AMPA receptors at hippocampal CA1 synapses and induction of NMDA receptor-independent long-term potentiation. Eur J Neurosci 13(8):1635–1643. doi:ejn1523 [pii]
39. Jeromin A, Yuan LL, Frick A et al (2003) A modified Sindbis vector for prolonged gene expression in neurons. J Neurophysiol 90(4):

2741–2745. doi:10.1152/jn.00464.2003 00464.2003 [pii]

40. Kim J, Dittgen T, Nimmerjahn A et al (2004) Sindbis vector SINrep(nsP2S726): a tool for rapid heterologous expression with attenuated cytotoxicity in neurons. J Neurosci Methods 133(1–2): 81–90. doi:S0165027003003248 [pii]
41. Konno A, Honjo T, Uchida A et al (2011) Evaluation of a Sindbis virus vector displaying an immunoglobulin-binding domain: antibody-dependent infection of neurons in living mice. Neurosci Res 71(4):328–334. doi:S0168-0102(11)02066-9 [pii] 10.1016/j.neures.2011.08.013
42. Ylosmaki E, Martikainen M, Hinkkanen A, Saksela K (2013) Attenuation of Semliki forest virus neurovirulence by microrna-mediated detargeting. J Virol 87(1):335–344. doi:JVI.01940-12 [pii] 10.1128/JVI.01940-12
43. Gervasi N, Hepp R, Tricoire L et al (2007) Dynamics of protein kinase A signaling at the membrane, in the cytosol, and in the nucleus of neurons in mouse brain slices. J Neurosci 27(11):2744–2750. doi:27/11/2744 [pii] 10.1523/JNEUROSCI.5352-06.2007
44. Hepp R, Tricoire L, Hu E et al (2007) Phosphodiesterase type 2 and the homeostasis of cyclic GMP in living thalamic neurons. J Neurochem 102(6):1875–1886. doi:JNC4657 [pii] 10.1111/j.1471-4159.2007.04657.x
45. Hu E, Demmou L, Cauli B et al (2011) VIP, CRF, and PACAP act at distinct receptors to elicit different cAMP/PKA dynamics in the neocortex. Cereb Cortex 21(3):708–718. doi:bhq143 [pii] 10.1093/cercor/bhq143
46. Ehrengruber MU, Lundstrom K (2007) Alphaviruses: Semliki forest virus and Sindbis virus vectors for gene transfer into neurons. Curr Protoc Neurosci Chapter 4:Unit 4:22. doi:10.1002/0471142301.ns0422s41
47. Davie JT, Kole MH, Letzkus JJ et al (2006) Dendritic patch-clamp recording. Nat Protoc 1(3):1235–1247. doi:nprot.2006.164 [pii] 10.1038/nprot.2006.164
48. Debanne D, Boudkkazi S, Campanac E et al (2008) Paired-recordings from synaptically coupled cortical and hippocampal neurons in acute and cultured brain slices. Nat Protoc 3(10):1559–1568. doi:nprot.2008.147 [pii] 10.1038/nprot.2008.147

Chapter 4

Gaussia Luciferase-Based Mycoplasma Detection Assay in Mammalian Cell Culture

M. Hannah Degeling, M. Sarah S. Bovenberg, Marie Tannous, and Bakhos A. Tannous

Abstract

Mycoplasma contamination in mammalian cell culture is a common problem with serious consequences on experimental data, and yet many laboratories fail to perform regular testing. In this chapter, we describe a simple and sensitive mycoplasma detection assay based on the bioluminescent properties of the *Gaussia* luciferase reporter.

Key words Mycoplasma contamination, Mycoplasma detection assay, Gluc mycosensor, Cell culture contamination, Antibiotics, Bioluminescence reaction

1 Introduction

Mycoplasmas are common contaminants of mammalian cell culture; due to their small size and the absence of a rigid cell wall, they have the ability to pass through most bacterial filters. In contrast to other contaminants such as fungi and bacteria, mycoplasma contamination is impossible to detect by eye, since the contamination does not influence the color of the cell culture medium, the pH, the turbidity, nor the odor. Mycoplasmas are also not visible under the regular light microscopes in cell culture laboratories. Furthermore, the fact that mycoplasma can grow both intracellular and extracellular greatly contributes to their resistance to several antibiotics in the extracellular cell culture medium, such as penicillin and streptomycin [1]. Due to these issues, mycoplasma contamination rates in mammalian cell culture laboratories have been found to be as high as 70 % [1–3].

Mycoplasma contamination can have a tremendous effect on host cells. For instance, mycoplasma can interfere with numerous

Authors Hannah Degeling and Sarah S. Bovenberg have contributed equally to this chapter.

Christian E. Badr (ed.), *Bioluminescent Imaging: Methods and Protocols*, Methods in Molecular Biology, vol. 1098, DOI 10.1007/978-1-62703-718-1_4, © Springer Science+Business Media New York 2014

cellular processes including cell metabolism, proliferation, gene expression, and function [2–7]. Therefore, all cell culture laboratories should test cell cultures for mycoplasma contamination on a regular basis. The ideal detection method for mycoplasma contamination should be simple to perform, sensitive, specific, rapid, inexpensive, and suitable for testing of numerous cell cultures simultaneously.

In this chapter, we describe a simple mycoplasma detection assay based on degradation of the *Gaussia* luciferase (Gluc) reporter protein. Typically, in conditioned medium of cells under regular culture conditions, Gluc has a half-life of >7 days [8]. The half-life of Gluc is tremendously decreased in the presence of mycoplasma contamination, and therefore the level of decline in Gluc activity correlates to mycoplasma infection rate [9]. This mycoplasma-specific decrease has been confirmed by several assays including Western blot analysis, which showed Gluc protein degradation over short period of time (2–24 h) only in the presence of mycoplasma contamination. We reasoned that this phenomenon could be used as a sensitive and specific biosensor to monitor the mycoplasma contamination in mammalian cells (Mycosensor) [9].

Around 90–95 % of all mycoplasma contamination in mammalian cell cultures is caused by either *M. orale*, *M. hyorhinis*, *M. arginini*, *M. fermentans*, *M. hominis*, or *A. laidlawii* [1]. The Gluc mycosensor has been confirmed on three of these commonly isolated mycoplasma strains including *Mycoplasma fermentens*, *Mycoplasma hominis*, and *Mycoplasma orale*. Importantly, this Mycosensor showed to be more sensitive in detecting mycoplasma contamination as compared to a commercially available bioluminescent-based assay and is amenable to high-throughput applications [9]. Alternative Mycoplasma detection methods are available including mycoplasma DNA amplification by polymerase chain reaction, which is sensitive but prone to errors; by traditional bacterial cell culture, which is very time consuming [10]; or by commercially available kits which have low sensitivity and are costly. The Gluc Mycosensor, however, is easy to use, sensitive, and especially suited for testing of numerous cell quantities simultaneously. Nevertheless, we recommend in all cases to combine two different detection methods to achieve and maintain total mycoplasma clearance in the cell culture hoods.

2 Materials

2.1 Gluc Recombinant Protein

1. Gluc recombinant protein (Nanolight).

   Or, for the preparation of Gluc recombinant protein:

2. cDNA encoding Gluc (Nanolight), and N-terminal pelB periplasmic signal sequence, and a C-terminal 6-His tag in pET26b(+) vector (Novagen).

3. HMS174 competent bacterial cells and LB containing 30 μg/mL kanamycin.
4. IPTG Bugbuster Master Mix containing Benzonase (Novagen).
5. NJ45 μm syringe filter nickel charged resin column (Novagen).
6. Binding buffer: 0.5 M NaCl, 20 mM Tris–HCl, 5 mM imidazole, pH 7.9.
7. Wash buffer: 0.5 M NaCl, 20 mM Tris–HCl, 60 mM imidazole, pH 7.9.
8. Elution buffer: 0.5 M NaCl, 20 mM Tris–HCl, 1 M imidazole, pH 7.9.
9. Coelenterazine and a luminometer.
10. 3,500 MW cut-off dialysis tubing (Fisher Scientific).
11. Bradford assay.
12. SDS-PAGE, NuPAGE„¥ 10 % Bis–Tris gel and coomassie blue staining (Invitrogen).

2.2 Gluc-Containing Medium

1. 293 T human fibroblast cells.
2. Lentivirus vector expressing Gluc and the enhanced green fluorescent protein (GFP) under the control of CMV promoter.
3. Polybrene.

2.3 Cell Culture

1. High glucose Dulbecco's modified Eagle's medium. Supplemented with 10 % fetal bovine serum, and 100 U/mL penicillin and 100 μg/mL streptomycin (*see* **Notes 1** and **2**).
2. Incubator with a humidified atmosphere supplemented with 5 % CO_2 at 37 °C.
3. 24-Well cell culture plates.

2.4 Bioluminescence Reaction

1. A luminometer to measure the bioluminescence activity of Gluc (*see* **Note 3**).
2. 20 μM coelenterazine (Nanolight). Diluted in phosphate buffer saline (PBS; *see* **Note 4**).
3. Black 96-well microtiter plates (*see* **Note 5**).

3 Methods

Be aware that all materials used for cell culture and the Mycosensor assay should be stored and used under sterile conditions to avoid contamination during the Gluc incubation period. An overview of the complete Gluc Mycosensor assay protocol is presented in Fig. 1.

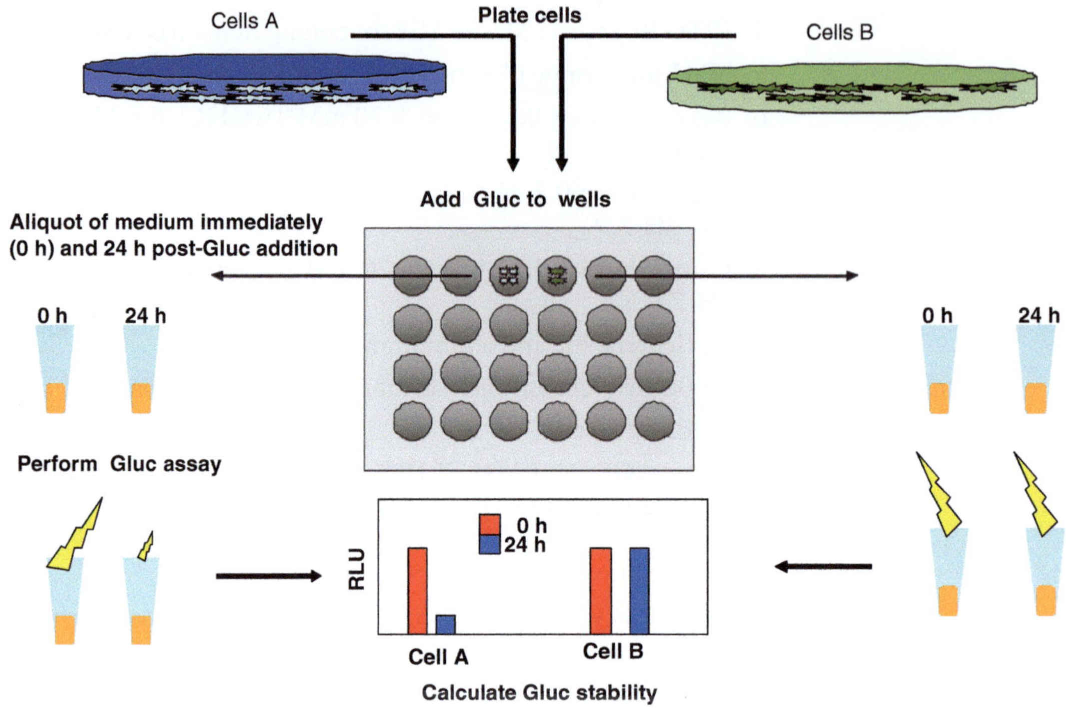

Fig. 1 Overview of the Mycosensor assay

3.1 Gluc Recombinant Protein Purification

In brief, the cDNA encoding Gluc is amplified by PCR using specific primers and cloned in-frame between an N-terminal pelB periplasmic signal sequence and a C-terminal 6-His tag in pET26b(+) vector using *Bam*HI and *Xho*I sites. In this vector, gene expression is under the control of an isopropyl β-d-1-thiogalactopyranoside (IPTG)-inducible T7 RNA polymerase promoter. The vector is transformed into HMS174 competent bacterial cells and grown overnight in 200 mL LB containing 30 μg/mL kanamycin. Once the culture reached an OD_{600} of 0.6, protein expression is induced by adding 20 μM IPTG to the culture which is grown for 18 h at room temperature. Cells should be pelleted by centrifugation for 15 min at 10,000 × *g* and resuspended in 10 mL Bugbuster Master Mix containing Benzonase. Insoluble debris is pelleted by another spin. The clarified lysate is then filtered through a 45 μm syringe filter and loaded onto a nickel charged resin column equilibrated with binding buffer. The column should be rinsed with 20× volumes of binding buffer followed by 18× volumes of wash buffer. His-tagged Gluc is eluted from the column with 1.2 mL of elution buffer collecting 200 μL fractions. The fraction containing the highest Gluc activity (as measured using coelenterazine and a luminometer) should be pooled and dialyzed against 30 mM

Tris–HCl, pH 8.0 overnight using 3,500 MW cut-off dialysis tubing. Glycerol is added to a final concentration of 10 % and protein concentration is determined using the Bradford assay. Purity of protein can be analyzed by SDS-PAGE under reducing conditions using a NuPAGE„¥ 10 % Bis–Tris gel and coomassie blue staining.

3.2 Gluc-Containing Medium

As an alternative to recombinant Gluc, conditioned medium from mycoplasma-free cells stably expressing Gluc can be aliquoted and stored at −80 °C and can be used for the Mycosensor assay. In this chapter, we take the widely used 293 T human fibroblast cells as an example for the Mycosensor assay. 239 T cells can be engineered to stably express the naturally secreted Gluc (*see* **Note 1**) by infecting these cells with a lentivirus vector expressing Gluc and the enhanced GFP under the control of CMV promoter at a multiplicity of infection of 10 transducing units/cell in the presence of 10 μg/mL Polybrene [8]. The next day, cells are washed and maintained in regular growth medium. GFP can be used as a marker to monitor transduction efficiency and therefore Gluc expression.

3.3 Mycosensor Assay

1. Culture the test cells for at least 72 h under normal culture conditions as described above (*see* **Note 2**).
2. Plate cells to be tested in triplicates in a 24-well plate at around 30 % confluency; an ideal amount of 293 T cells would be 1×10^5 cells/well, however, as low as 5×10^4 cells can be used. Cells should be plated in 500 μL of growth medium (*see* **Note 6**). In order to save time, it is also possible to simply assay the cell-free conditioned medium of the test cells by transferring a small aliquot to the 24-well plate instead of replating the cells. In this case, the Gluc-containing medium or purified Gluc can be added immediately and **step 3** can be skipped. One should take into consideration that this technique is much less sensitive than the standard protocol (see **Note 7**).
3. The next day, add purified Gluc to a final concentration of 35 ng/mL to the conditioned medium. Alternatively, if one chooses to use Gluc-containing medium, remove 250 μL of medium from cells to be tested and add 250 μL of Gluc-containing media (*see* **Note 8**).
4. Mix by stirring the plate and immediately take a 50 μL aliquot from each well; this will be used as the time point zero. Samples can be frozen at −80 °C until further analysis (*see* **Note 9**).
5. Return the plate to the cell culture incubator.
6. At 24 h post-Gluc addition, take another 50 μL aliquot from each well and freeze at −80 °C. If fast analysis is required, and if mycoplasma contamination is high, it is possible to detect a decrease in Gluc activity (and therefore mycoplasma contamination) after 4 h. However, if mycoplasma contamination is

low, 4 h will not yield a significant degradation of Gluc and longer incubation time is needed. One may choose to perform time analysis by storing an aliquot of conditioned medium at different time points.

7. Do not discard the cells at the 24 h time point since very low levels of mycoplasma can be detected after 48 or 72 h, even if test results are negative at earlier time points (*see* **Note 9**).
8. Assay these aliquots for Gluc activity using coelenterazine and a luminometer (*see* below).

3.4 The Mycosensor Bioluminescence Assay

9. Thaw aliquots from the different time points to room temperature (*see* **Note 10**).
10. While waiting, dilute coelenterazine in PBS to a final concentration of 20 μM and incubate for 20–30 min in the dark (*see* **Note 11**).
11. Transfer 20 μL of each aliquot into black 96-well microtiter plate.
12. Using a luminometer with a built-in injector, inject 80 μL of coelenterazine and acquire the signal immediately for 10 s (*see* **Note 12**).

3.5 The Mycosensor Assay Results

Mycoplasma contamination is determined by comparing the Gluc signal at time point B (e.g., 24 h) to the signal at time zero. A significant decrease in Gluc activity indicates mycoplasma contamination (*see* **Note 13**). Typical results from a positive and negative mycoplasma assay are presented in Fig. 2.

4 Notes

1. Be alert that the cells used to express Gluc, and from which the conditioned medium is used for the Mycosensor assay, are not contaminated with mycoplasma. We recommend to test these cells on a regular basis with two different mycoplasma detection assays such as: MycoAlert® (Lonza Rockland, Rockland, ME) and PCR PromoKine Mycoplasma Test KIT I/C (PromoCell, Heidelberg, Germany) according to the manufacturer's standard protocol.
2. Mycoplasma needs at least 48–72 h to recover from frozen stocks and therefore cells should be cultured for at least 3 days after thawing from liquid nitrogen before testing for contamination.
3. Any plate reader with a built-in injector that can adequately measure bioluminescence can be used.
4. Other concentrations of CZN can be used, up to 100 μM; however, the 20 μM concentration of CZN in combination

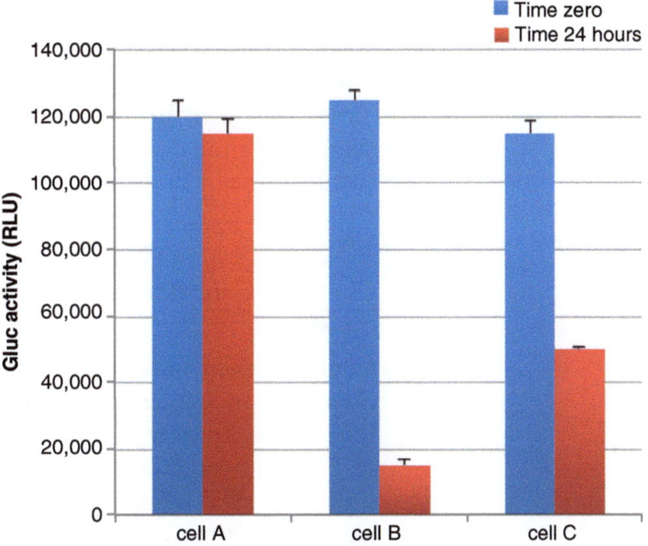

Fig. 2 Typical mycoplasma contamination results obtained using the Gluc Mycosensor

with the luminometer used in our laboratory, and the black microtiter plate will lead to the most optimum Gluc value under our assay conditions.

5. This note is also to some extent applicable to **Notes 3** and **4**; be aware that the Gluc signal can reach saturation depending on the amount of Gluc used and the sensitivity of the luminometer. In the case that Gluc-containing conditioned medium is used for the Mycosensor assay, this problem is more likely to occur. The use of black plates could solve this problem as they typically yield up to one log lower signal as compared to white plates. As an alternative, one can dilute the medium before the assay.
6. Keep in mind that plating too many cells will result in overconfluency at 72 h resulting in cell death and therefore the Mycosensor will not be accurate.
7. As an alternative to testing mycoplasma contamination on plated cells, recombinant Gluc can be added directly to an aliquot of conditioned medium from cells to be tested; however, this strategy is less sensitive as compared to testing the cells themselves. One may choose to do the assay on conditioned medium initially and if the results are negative, it can then be confirmed on plated cells.
8. Be careful when adding Gluc not to disrupt the cells, since some cell types are easily detached from the plate leading to false results.
9. It is of uttermost importance to include a time point zero when performing the Mycosensor at a range of time points.

It is recommended to take two different aliquots of conditioned medium at time point zero in case one chooses to run the 24 h time point first. If test results are negative, the 48 and 72 h time point can then be assayed and compared to the second aliquot of time point zero.

10. We observed that aliquots from the same sample but treated in a different manner (e.g., frozen versus not frozen) can show some variation in Gluc signal intensity. It is important to treat all aliquots, including the ones from different time points, in the same manner.
11. Coelenterazine purchased should be first reconstituted in acidified methanol (one drop of HCl to 10 mL of Methanol) to a final concentration of 5 mg/mL. Just before use, it is diluted in PBS to a final concentration of 20 μM. Coelenterazine is prone to auto-oxidation and a decrease in bioluminescence signal over time can simply be observed due to this artifact. Incubating diluted coelenterazine for 20 min in the dark stabilizes the signal and will result in a stable bioluminescent output.
12. A luminometer with built-in injector is required to perform the Mycosensor assay, since Gluc catalyzes a flash-type bioluminescence reaction. Therefore, the reaction starts with a high signal, followed by a rapid decrease over the course of few seconds [11]. If a luminometer with built-in injection is not available, one can dilute coelenterazine in PBS containing 0.1 % Triton X-100. We showed that this detergent yields some stability to the Gluc signal, especially when used in combination with GlucM43I mutant [12]. Alternatively, stabilization kits are available from Targeting Systems or Nanolight (GAR-reagents). The substrate should be diluted in the appropriate reagent and can be added manually to the medium and assayed using a luminometer.
13. We recommend testing cells for mycoplasma contamination on a regular basis using the Gluc Mycosensor. Confirm negative results with another type of assay, such as the PCR-based mycoplasma assay.

Acknowledgements

This work was supported by grant from NIH/NINDS P30NS045776 and 1R01NS064983 (BAT). M. Hannah Degeling is supported by a Fulbright scholarship, the Saal van Zwanenberg Foundation, VSB fonds, Dr. Hendrik Muller Vaderlandschfonds, the Dutch Cancer Foundation (KWF Kankerbestrijding), the Hersenstichting brain fund as well as the Jo Keur (Leiden hospital).

M. Sarah Bovenberg was supported by a Fulbright scholarship, the Huygens Scholarship Program, VSB fonds and the Saal van Zwanenberg Foundation. We thank Kristan van der Vos for assistance with cell culture.

References

1. Drexler HG, Uphoff CC (2002) Mycoplasma contamination of cell cultures: Incidence, sources, effects, detection, elimination, prevention. Cytotechnology 39(2):75–90
2. Hay RJ, Macy ML, Chen TR (1989) Mycoplasma infection of cultured cells. Nature 339(6224):487–488
3. Rottem S, Barile MF (1993) Beware of mycoplasmas. Trends Biotechnol 11(4): 143–151
4. Logunov DY, Scheblyakov DV, Zubkova OV et al (2008) Mycoplasma infection suppresses p53, activates NF-kappaB and cooperates with oncogenic Ras in rodent fibroblast transformation. Oncogene 27(33):4521–4531
5. Namiki K, Goodison S, Porvasnik S et al (2009) Persistent exposure to Mycoplasma induces malignant transformation of human prostate cells. PloS One 4(9):e6872
6. Zhang S, Tsai S, Wu TT et al (2004) Mycoplasma fermentans infection promotes immortalization of human peripheral blood mononuclear cells in culture. Blood 104(13): 4252–4259
7. Zinocker S, Wang MY, Gaustad P et al (2011) Mycoplasma contamination revisited: mesenchymal stromal cells harboring Mycoplasma hyorhinis potently inhibit lymphocyte proliferation in vitro. PloS One 6(1):e16005
8. Wurdinger T, Badr C, Pike L et al (2008) A secreted luciferase for ex vivo monitoring of in vivo processes. Nat Methods 5(2):171–173
9. Degeling MH, Maguire CA, Bovenberg MS, Tannous BA (2012) Sensitive assay for mycoplasma detection in mammalian cell culture. Anal Chem 84(9):4227–4232
10. Uphoff CC, Drexler HG (2002) Detection of mycoplasma in leukemia-lymphoma cell lines using polymerase chain reaction. Leukemia 16(2):289–293
11. Tannous BA, Kim DE, Fernandez JL et al (2005) Codon-optimized Gaussia luciferase cDNA for mammalian gene expression in culture and in vivo. Mol Ther 11(3):435–443
12. Maguire CA, Deliolanis NC, Pike L et al (2009) Gaussia luciferase variant for high-throughput functional screening applications. Anal Chem 81(16):7102–7106

Part II

Imaging Molecular and Cellular Events

Chapter 5

Split *Gaussia* Luciferase for Imaging Ligand–Receptor Binding

Kathryn E. Luker and Gary D. Luker

Abstract

Ligand binding to cell surface receptors activates signaling pathways in normal and pathologic conditions, and internalized ligand–receptor complexes may continue to signal from endosomes. Accessibility of cell surface receptors and the central function of ligand–receptor binding in signal transduction make ligand binding a prime target for therapeutic agents. We describe a *Gaussia* luciferase complementation method for imaging ligand–receptor binding in cell-based assays and living mice. While we illustrate this imaging method for chemokine ligand CXCL12 and its receptors CXCR4 and CXCR7, this imaging strategy can be generalized to a large number of ligand–receptor interactions.

Key words Molecular imaging, Optical imaging, Split luciferase, Bioluminescence, Protein complementation assay, PCA

1 Introduction

Protein fragment complementation assays based on luciferase enzymes, referred to as luciferase complementation or split luciferase assays, provide a powerful strategy to quantify protein–protein interactions in formats ranging from cell lysates to intact mice. Luciferase complementation entails fusing inactive amino (N)-terminal and carboxy (C)-terminal enzyme fragments to two different proteins of interest. Interactions between proteins of interest bring N- and C-terminal luciferase fragments together to reconstitute an active enzyme, producing bioluminescence as a quantitative measure of interactions between target proteins. Dissociation of proteins of interest also separates fused N-terminal and C-terminal luciferase fragments and reduces bioluminescence, allowing dynamic changes in protein–protein interactions to be quantified in real time. These assays may be used to quantify regulation of protein–protein complexes in response to signaling events, chemical probes, or drugs. When luciferase complementation reporters are expressed stably in mammalian cell lines, the same

Christian E. Badr (ed.), *Bioluminescent Imaging: Methods and Protocols*, Methods in Molecular Biology, vol. 1098,
DOI 10.1007/978-1-62703-718-1_5, © Springer Science+Business Media New York 2014

reporter cells can transition directly from intact cells to animal models, providing a facile approach for development and preclinical testing of candidate drugs.

Monitoring ligand–receptor binding places unique demands on a luciferase complementation assay system. These requirements include that the enzyme functions in the extracellular space and minimizes steric constraints imposed by fusing enzyme fragments to a ligand and receptor. Of available luciferase complementation assays, *Gaussia* luciferase (GLuc) best meets parameters for imaging ligand–receptor interactions [1]. GLuc does not require ATP, allowing enzyme activity in the extracellular space, and small size of N-terminal and C-terminal enzyme fragments (≈9–10 kDa) substantially reduces the potential to alter functions of fusion proteins. GLuc complementation is fully reversible, so ligand binding to a receptor and subsequent dissociation can be monitored in real time. We have used GLuc complementation to quantify binding of chemokine CXCL12 to its receptors CXCR4 and CXCR7 in intact cells and a mouse model of human breast cancer [2]. More generally, the GLuc complementation system is applicable to any ligand–receptor pair that can be modified to express N-terminal and C-terminal fragments of this enzyme.

2 Materials

2.1 Molecular Biology

1. DNA encoding open reading frames for desired interacting proteins.
2. Full-length *Gaussia* luciferase (GLuc) plasmid (New England Biolabs) or plasmids with NGLuc (amino acids 1–93) and CGLuc (amino acids 94–169) fragments.
3. Expression vectors with constitutive promoters for expression in mammalian cells.
4. Expression vector and packaging constructs for producing lentiviral vectors (optional).
5. Enzymes, buffers, and equipment for PCR, restriction digests of DNA, and ligations

2.2 Cell Culture

1. HEK-293 T cells or other cell line that can be transfected readily.
2. Desired cell line(s) for biologic question of interest.
3. General supplies for cell culture, including media, plasticware, and incubators.

2.3 Cell Imaging

1. 96-Well plates with black sides, clear bottom, and lid for tissue culture.
2. Multichannel pipets for volumes from 1 to 200 μl.

3. Sterile pipette tips with low adherence coating adherence.
4. Sterile 1× phosphate-buffered saline (PBS) solution.
5. Stock solution of coelenterazine (Promega or other vendor) 1 mg/ml in methanol, stored in tightly closed container at −20 °C (*Gaussia* luciferase substrate).
6. Acid wash solution: 0.2 M acetic acid, 0.5 M NaCl (optional) (*see* **Note 1**).
7. Bioluminescence imaging system with high sensitivity and software for data quantification and analysis (IVIS, Perkin-Elmer; or similar system).

2.4 Animal Imaging

1. Appropriate mouse strain for desired experimental system, such as immunocompromised mice (nude, SCID, or NSG) for human tumor xenografts.
2. Small animal shaver such as Wahl compact cordless trimmer (optional).
3. Depilatory solution such as Nair (optional).
4. 10 mg/ml coelenterazine stock in acidified methanol, store in tightly sealed container at–20 °C (*see* **Note 2**).
5. Sterile solution 40 % DMSO in PBS for diluting coelenterazine immediately before injection and imaging.
6. 28–30 gauge insulin syringe for intravenous tail vein injection in mice.
7. Restraint device for tail vein injections (Braintree Scientific Tail Vein Injection Platform or other similar device) (optional).
8. Bioluminescence imaging instrument with isoflurane anesthesia (IVIS or similar instrument as described in Subheading 2.3).

3 Methods

3.1 Construct Gaussia Luciferase Complementation Reporters

1. Select a ligand and corresponding receptor as interacting proteins and determine positions of NGLuc and CGLuc fusions to these proteins (Fig. 1) (*see* **Note 3**).
2. Design and optimize linkers between GLuc enzyme fragments and respective ligand and receptor pairs. While not required, linkers may limit steric constraints on ligand–receptor binding and folding of GLuc fragments (*see* **Note 4**).
3. Generate fusion proteins for ligand and receptor pairs using appropriate molecular biology procedures. We typically produce all logical orientations of fusions with NGLuc and CGLuc (*see* **Note 5**).
4. Generate relevant control constructs for nonspecific association of NGLuc and CGLuc (*see* **Note 6**).

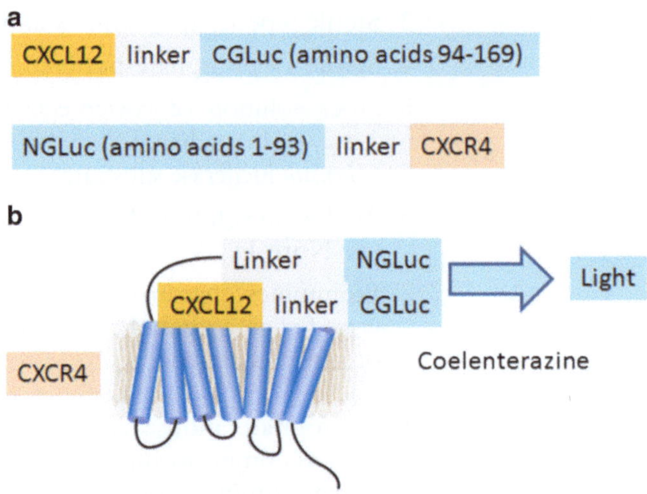

Fig. 1 Schematic diagrams of GLuc complementation constructs and ligand–receptor interaction illustrated for CXCL12 and CXCR4. (**a**) Complementation reporter constructs with chemokine CXCL12 fused via a linker to CGLuc and NGLuc fused with an intervening linker to seven-transmembrane chemokine receptor CXCR4. These positions of NGLuc and CGLuc enzyme fragments allow complementation to occur in both extracellular and intracellular compartments. We also tested constructs with CXCL12 fused to NGLuc and CGLuc fused to CXCR4, but these reporters produced less bioluminescence upon ligand–receptor binding. (**b**) Binding of CXCL12-CGLuc to NGLuc-CXCR4 reconstitutes GLuc activity to oxidize the substrate coelenterazine and produce bioluminescence (figure adapted from ref. 10)

5. Express complementation reporters in appropriate vectors for mammalian cells. Vectors should be selected with markers, such as co-expressed fluorescent proteins or antibiotic resistance genes, suitable for generating stable cell lines (*see* **Note 7**).
6. Confirm complementation constructs by DNA sequencing.

3.2 Cell-Based Bioluminescence Imaging

1. We initially test pairs of NGLuc and CGLuc fusions with ligand and receptor by transient transfection in 293T cells or another cell line that transfects readily. We include appropriate control NGLuc and CGLuc constructs in these tests. The purpose of these initial tests is to identify an optimal pair of NGLuc and CGLuc fusions for use in stable cell lines and subsequent cell-based assays and living mice. The format for cell-based assays is the same for transiently transfected cells or cells stably expressing reporter constructs (*see* **Note 8**).
2. Plate cells in black-walled, clear bottom 96-well plates for tissue culture. Cell density should be 1×10^4–2×10^4 cells/per well in 100 μl complete growth medium with serum. Culture cells overnight under standard conditions in preparation for assays the following day (*see* **Note 9**).

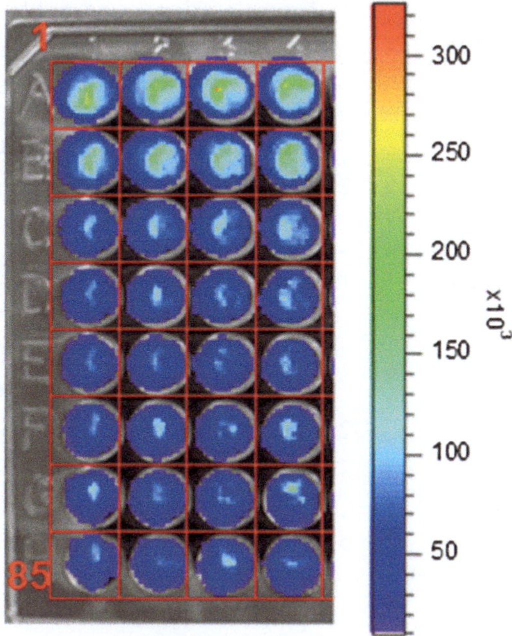

Fig. 2 Cell-based GLuc complementation assay for inhibition of CXCL12-CGLuc binding to NGLuc-CXCR4. Equal numbers (1×10^4 cells each) of MDA-MB-231 cells secreting CXCL12-CGLuc (231-CXC12-CGLuc) or expressing NGLuc-CXCR4 (231-NGLuc-CXCR4) were cocultured in black wall 96 plates. Cells were incubated with vehicle control (*row A*) or increasing concentrations of the CXCL12-CXCR4 inhibitor AMD3100, which decreased bioluminescence from GLuc complementation. Grid overlay is used for ROI analysis. Scale bar shows range of photon flux values by pseudocolor display with *red* being highest and *blue* lowest numbers

3. Remove cell culture medium and replace with a minimum volume of fresh phenol red DMEM medium for assays (30–40 μl). For assays using inhibitors, we prepare 10× stocks of desired dilutions so that inhibitor or vehicle control can be added in a small volume, such as adding 4 μl of 10× inhibitor stock to 36 μl of medium per well. To perform time course studies, we remove standard culture medium and replace with phenol red free DMEM at staggered times so all wells in the plate are imaged at the end (*see* **Note 10**).
4. Measure *Gaussia* luciferase complementation signal by adding a 1:500 dilution of 1 mg/ml coelenterazine stock in phenol red free medium to each well using a multichannel pipette (*see* **Note 11**).
5. Image plate on the bioluminescence instrument as soon as possible after adding coelenterazine. A typical image requires 30–60 s with large binning (*see* **Note 12**).
6. Quantify bioluminescence by region-of-interest (ROI) analysis using software on the bioluminescence imaging instrument (Fig. 2).

3.3 Construction of Mouse Tumor Model

1. Implant mixture of stable complementation reporter cells for ligand and receptor (≈0.5–2 × 10^6 cells total) subcutaneously or orthotopically, such as in the mammary fat pad, in appropriate strain of mouse. We typically inject a 1:1 mixture of ligand and receptor complementation cells (*see* **Note 13**).
2. Begin imaging experiments when 4–5 mm diameter tumors form (*see* **Note 14**).

3.4 Mouse Imaging

1. Prepare coelenterazine for intravenous injection into one mouse by adding 44 μl of 10 mg/ml coelenterazine stock to 66 μl of 40 % DMSO/PBS solution. We use 100 μl of this solution for tail vein injection in each mouse (*see* **Note 15**).
2. Anesthetize mice with 1–2 % isoflurane and maintain mice under anesthesia during tail vein injection of coelenterazine (*see* **Note 16**).
3. Transfer mouse immediately to bioluminescence instrument and acquire image (*see* **Note 17**).
4. Remove mouse from imaging instrument and monitor for complete recovery from anesthesia.
5. Quantify imaging data by region-of-interest (ROI) analysis of bioluminescence produced by the tumor site, using units of photon flux (Fig. 3) (*see* **Note 18**).

4 Notes

1. Acid washing dissociates ligand–receptor complexes in the extracellular space, allowing quantification only of internalized receptors with bound ligand. This procedure is explained further in **Note 11**.
2. Acidified methanol is needed to keep higher concentrations of coelenterazine in solution. We dissolve 10 mg coelenterazine in 1 ml final volume of methanol and 3 N HCl (980 μl of methanol and 20 μl of 3 N HCl).
3. NGLuc and CGLuc must both localize to the extracellular space to detect ligand–receptor binding at the cell surface. To accomplish this objective, a GLuc fragment should be attached to the extracellular terminus of a transmembrane receptor and the terminus of the ligand that does not primarily determine receptor binding. We use the GLuc enzyme fragments identified by Remy et al. in which the enzyme is divided between amino acids G93 and E94 (omitting the 16 amino acid leader sequence) to form NGLuc and CGLuc with amino acids 1–93 and 94–169 [1].
4. We typically add a flexible linker, such as amino acids GGGSGGGS, between GLuc enzyme fragments and proteins

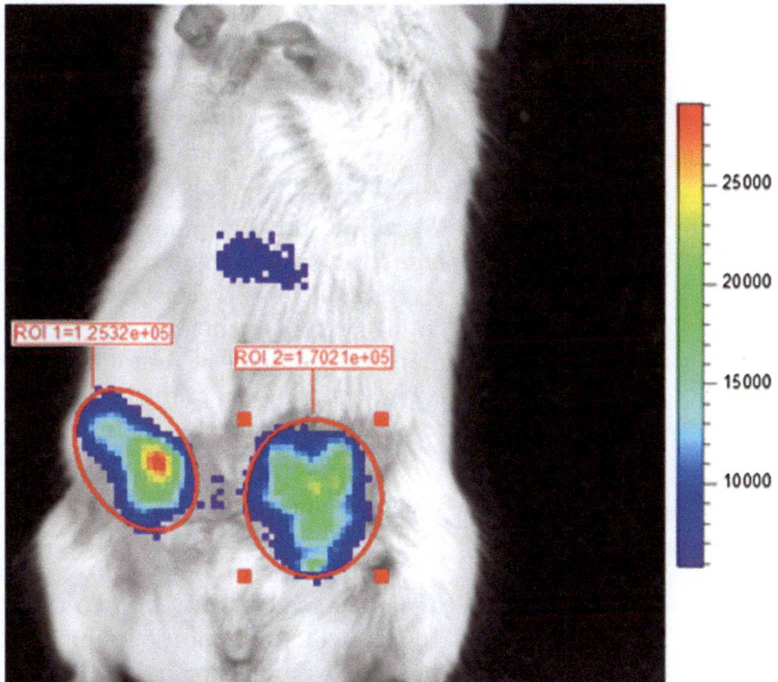

Fig. 3 Imaging ligand–receptor binding in living mice. Mice were implanted with equal numbers of 231-CXCL12-CGLuc and 231-NGLuc-CXCR4 cells as orthotopic mammary tumor xenografts in NSG mice. Imaging began 20 s after tail vein injection of coelenterazine, using 2 min acquisition and large binning. Circles and values show photon flux measurements for ROIs around each tumor. Scale bar shows range of values depicted by pseudocolor display

of interest. Linkers allow placement of restriction enzyme sites for cloning fusion constructs and potentially reduce steric constraints on ligand–receptor pairs and GLuc enzyme fragments. A linker sequence is not required, and users should consider testing fusion constructs with shorter or no linking amino acids to determine optimum *Gaussia* luciferase complementation output for the selected ligand–receptor pair.

5. We test NGLuc and CGLuc fusions to both ligand and receptor in all orientations that allow complementation in the extracellular space with the goal of identifying a combination that optimizes ligand-dependent bioluminescence relative to background levels.
6. Control constructs could have mutations in key amino acids in either ligand or receptor that confer specific binding, secreted and/or membrane bound extracellular NGLuc or CGLuc fragments, or a mismatched pair of ligand and receptor.
7. We typically generate stable cell lines via lentiviral vectors with co-expressed fluorescent proteins, allowing us to use flow

cytometry to sort for batch populations of transduced cells. Alternatively, investigators may transfect standard expression constructs into cells and select for cells with stable expression of the reporter transgene by drug resistance or fluorescence. We prefer batch populations of cells to avoid potential confounding effects of clonal cell lines.

8. We select the optimum reporter pair based on maximum induction of bioluminescence produced by matched ligand–receptor fusion proteins above background levels defined by control complementation reporters. After identifying the optimum orientations of fusion proteins, we generate stable cell lines expressing either the ligand or receptor complementation reporters, respectively.

9. We typically plate equal numbers of cells that either secrete a GLuc complementation ligand or express the cognate GLuc complementation receptor to make a total of 1×10^4–2×10^4 cells/well. These cocultures reproduce chronic intercellular signaling, such as between two different cell types in a tumor. To test acute induction of complementation signal for a soluble ligand binding to its receptor, users can collect supernatants from ligand-secreting cells and add the supernatant to cells expressing the complementation receptor.

10. We use phenol red free medium to improve transmission of GLuc bioluminescence from cells. Since serum oxidizes coelenterazine, the substrate for GLuc, and generates substantial background signal, we routinely use serum-free medium for assays. We have performed complementation assays in serum-free medium for up to 24 h with no loss of signal. Users may need to tailor lengths of experiments and culture conditions based on specific cell types. If a low percentage of serum is needed to maintain cell viability for extended assays, users should make certain to have control cells with the same percentage of serum to determine background bioluminescence.

11. GLuc bioluminescence decreases by approximately 70 % within 1 min of adding coelenterazine [3], so we add coelenterazine to wells as rapidly as possible and begin imaging as soon as possible thereafter. Adding coelenterazine to medium already present in wells measures total ligand–receptor complexes in extracellular and intracellular compartments. To quantify only internalized, intact ligand–receptor complexes, users should use an acid wash protocol to dissociate extracellular ligand–receptor pairs. The acid wash procedure entails (a) removing all medium from wells; (b) incubating cells on ice with 150 μl per well ice-cold acid wash solution for 3–5 min; (c) removing acid wash; (d) washing wells once with 200 μl per well warm PBS; and (e) adding 1:500 coelenterazine diluted in PBS. Using PBS further decreases background bioluminescence as compared with phenol red free medium.

12. Imaging times need to be optimized for specific pairs of ligand and receptor to acquire detectable signal without saturating the detector system. If the bioluminescence signal is low, users may increase numbers of cells per well, use more concentrated coelenterazine (1:100 to 1:250 dilution of 1 mg/ml stock) or acquire images for longer periods of time.
13. Tumor xenografts provide a localized environment in which large numbers of ligand and receptor complementation pairs exist in a confined, relatively small volume. Adding Matrigel to implanted tumors may improve tumor take and generate more well-defined tumors. The configuration of an implanted tumor favors a large number of ligand–receptor binding events, enhancing detection of GLuc complementation. Since overlying tissue substantially attenuates blue light produced by GLuc, tumors in a superficial site (such as mammary fat pad or subcutaneous implant) produce a greater imaging signal than tumors in internal organs. Initial animal imaging studies should include a control group of mice with a control complementation pair that should not interact specifically, allowing quantification of background signal. Cancer cells have been shown to migrate from one tumor to another when two tumors are implanted in the same mouse, so we prefer to have separate cohorts of animals for ligand–receptor and control groups [4].
14. If the complementation signal is particularly weak, larger tumors may be needed to produce a detectable imaging signal. Tumor burden must comply with protocols approved by institutional committees for care and use of animals.
15. Coelenterazine oxidizes in aqueous solutions, so we prepare the diluted volume of coelenterazine in 40 % DMSO/PBS immediately before tail vein injection into each mouse. Larger amounts of coelenterazine typically are needed to detect GLuc complementation relative to full-length, intact enzyme since the complemented enzyme fragments generate only ≈25 % as much light. Adjust amounts of injected coelenterazine as needed to produce a detectable signal in mice.
16. We prefer isoflurane anesthesia because of its rapid induction, ease of adjusting dosage, and rapid recovery relative to injectable anesthetics. Anesthetizing mice prior to tail vein injection allows the mouse to be transferred immediately to the bioluminescence imaging instrument. Since GLuc has flash kinetics of bioluminescence, a delay between injection of coelenterazine and imaging results in loss of signal.
17. Image acquisition times vary depending on complementation signal intrinsic to the ligand–receptor pair, size of tumor, and amount of injected coelenterazine. Imaging times typically are 1–3 min per mouse, but times will need to be optimized for different ligand–receptor pairs and tumor sites.

18. IVIS and similar bioluminescence imaging instruments have dedicated software for defining ROIs and measuring photon flux. As tumors grow larger, investigators may need to acquire images for shorter periods of time to avoid saturating the detector system. Photon flux measurements correct for differences in imaging time used in different studies over the course of an experiment. When comparing bioluminescence for several mice in experimental and control groups, investigators should select the same minimum and maximum values for pseudocolor display prior to defining ROIs that encompass light emitted from a tumor or other site. This approach maintains consistency of data across different groups of mice, such as those treated with a specific inhibitor of ligand–receptor binding or vehicle control.

    To account for effects of tumor size and relative numbers of cells on GLuc complementation signal in different mice, we typically co-express a second, constitutive reporter protein in each complementation cell line. For example, beetle luciferases such as firefly or click beetle red can be used as a marker protein in one cell population. Bioluminescence from these enzymes can be distinguished readily from GLuc because beetle luciferases use a different substrate, luciferin. Beetle luciferases can be imaged immediately following GLuc using previously described protocols for these enzymes in animal studies [5]. Typically, bioluminescence from beetle luciferases is substantially greater than GLuc complementation, particularly since the complementation signal decreases rapidly. If necessary, using an emission filter of 580–600 nm will further discriminate light from relatively red shifted beetle luciferases from GLuc bioluminescence. To mark a second population of complementation cells in a tumor, we have used a far red fluorescent protein such as eqFP650 [6], which can be detected readily in subcutaneous sites such as mammary fat pads. Since near-infrared light transmits through tissues even better than far red, the recently described IFP may allow even more sensitive detection of cells in vivo [7]. These independent markers allow changes in ligand–receptor binding and GLuc complementation to be normalized to changes in ligand-producing and receptor cell lines overtime.

    Bioluminescence from GLuc peaks at 480 nm, so light from this enzyme is particularly sensitive to attenuation by overlying tissue, pigment, and fur [8]. To optimize detection of GLuc complementation in vivo, we typically shave fur overlying a tumor site and then use a depilatory agent to remove remaining hair. A depilatory agent should be applied briefly (<2 min), using a cotton swab or soft tissue to remove the lotion and fur. The skin site should be washed thoroughly with warm water to avoid chemical burns.

Coelenterazine is a substrate for the multidrug transporter ABCB1 (MDR1 P-glycoprotein), which limits access of this compound to the intact central nervous system because of ABCB1 in the blood–brain barrier [9]. ABCB1 also may limit intracellular accumulation of substrate and detection of internalized ligand–receptor complexes in cells that express this transporter. Investigators may need to test for expression of ABCB1 by flow cytometry or other assays and/or select an alternative cell line if needed to overcome this limitation.

Acknowledgement

This work was supported by NIH Grants R01CA136553, R01CA136829, R01CA142750, and P50CA093990.

References

1. Remy I, Michnick S (2006) A highly sensitive protein-protein interaction assay based on *Gaussia* luciferase. Nat Methods 3:977–979
2. Luker K, Mihalko L, Schmidt B, Lewin S, Ray P et al (2012) *In vivo* imaging of ligand receptor binding with *Gaussia* luciferase complementation. Nat Med 18:172–177
3. Tannous B, Kim D, Fernandez J, Weissleder R, Breakefield X (2005) Codon-optimized *Gaussia* luciferase cDNA for mammalian gene expression in culture and in vivo. Mol Ther 11:435–443
4. Kim M, Oskarsson T, Acharyya S, Nguyen D, Zhang X et al (2009) Tumor self-seeding by circulating cancer cells. Cell 139:1315–1326
5. Luker G, Luker K (2011) Luciferase protein complementation assays for bioluminescence imaging of cells and mice. Methods Mol Biol 680:29–43
6. Shcherbo D, Shemiakina I, Ryabova A, Luker K, Schmidt B et al (2010) Near infrared fluorescent proteins. Nat Methods 7:827–829
7. Filonov G, Piatkevich K, Ting L, Zhang J, Kim K et al (2011) Bright and stable near-infrared fluorescent protein for in vivo imaging. Nat Biotechnol 29:757–761
8. Weissleder R, Ntziachristos V (2003) Shedding light onto live molecular targets. Nat Med 9:123–128
9. Pichler A, Prior J, Piwnica-Worms D (2004) Imaging reversal of multidrug resistance in living mice with bioluminescence: MDR1 P-glycoprotein transports coelenterazine. Proc Natl Acad Sci U S A 101:1702–1707
10. McGarrigle D, Huang X-Y (2007) GPCRs signaling directly through Src-family kinases. Sci STKE 2007:pe35

Chapter 6

Video-Rate Bioluminescence Imaging of Protein Secretion from a Living Cell

Takahiro Suzuki and Satoshi Inouye

Abstract

We have established a method of video-rate bioluminescence imaging to visualize exocytotic protein secretion from a single living cell using *Gaussia* luciferase (GLase) as a reporter protein. The luminescence signals produced by the enzymatic reaction of secreted GLase (luciferase) and coelenterazine (luciferin) are detected with an electron-multiplying charge-coupled device camera. An exocytotic event of protein secretion can be visualized using the protein fused to GLase with a time resolution of 30–500 ms. Signal analyses of the bioluminescence video images reveal a number of exocytotic sites, the frequency of exocytotic events, and the amount of secreted protein on a whole cell. Furthermore, the method can distinguish between secreted and cell surface-bound proteins. Our method is a direct approach to investigate the secretion and localization of proteins on the whole surface of living cells.

Key words Video-rate bioluminescence imaging, *Gaussia* luciferase, Protein secretion, Exocytosis, Cell surface

1 Introduction

Real-time visualization of protein trafficking and distribution in living cells can provide insights into biological functions of proteins. Bioluminescence imaging of protein secretion from a single living cell has been carried out on the basis of a luciferin-luciferase reaction [1–5]. In mammalian cells, the secretory proteins including peptide hormones are transported by the secretory vesicles and released from cells. To investigate the exocytotic events on the cell surface, we recently developed a method of video-rate bioluminescence imaging using secreted *Gaussia* luciferase (GLase) as a reporter protein and an electron-multiplying charge-coupled device (EM-CCD) camera as a sensitive detector [4, 5]. Specifically, the secretory protein of interest is fused to GLase and expressed in mammalian cells, and exocytotic events from a whole cell are visualized (Fig. 1a, b). In our imaging system, the fused proteins of GLase to prepro-forms of insulin [4] and matrix metalloproteinase-2

Christian E. Badr (ed.), *Bioluminescent Imaging: Methods and Protocols*, Methods in Molecular Biology, vol. 1098,
DOI 10.1007/978-1-62703-718-1_6,

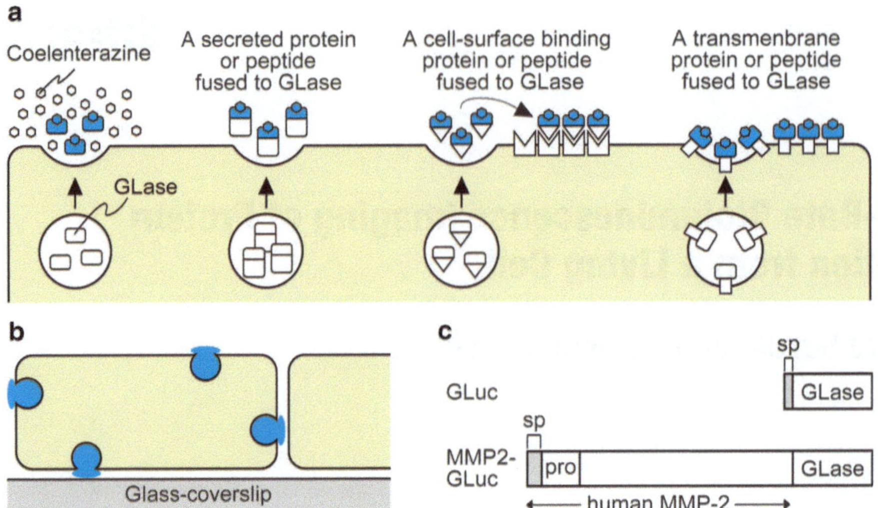

Fig. 1 Bioluminescence imaging of GLase as a reporter protein to visualize exocytotic protein secretion on the surface of mammalian cells. (**a**) Schematic representation of bioluminescence imaging of GLase and the fusion protein of GLase secreted from mammalian cells and its binding on the cell surface. (**b**) Detectable area of exocytosis in a single mammalian cell with bioluminescence imaging using secreted GLase. (**c**) Schematic representation of expression vectors for GLase and MMP2-GLase. GLuc (pcDNA3-GLuc); GLase with the signal peptide sequence, MMP2-GLuc (pcDNA3-hMMP2-GLuc); human MMP-2 preproprotein fused to GLase. Reproduced in part from Suzuki et al. [5] with the author's copyright under *PLoS One* editorial policies

(MMP-2) [5] were expressed in mammalian cells, and the exocytotic secretion of these fusion proteins from single cells was clearly visualized as luminescence spots. These results suggested that the method of video-rate bioluminescence imaging could be applied to investigate the protein secretion and the localization of various types of proteins on the whole surface of a living cell (Fig. 1a).

Here, we describe the application of video-rate bioluminescence imaging and MMP-2 secretion from HeLa cells using the fusion protein of MMP-2 to GLase (MMP2-GLase).

1.1 GLase and EM-CCD Camera

Secreted GLase is the smallest luciferase (consisting of 168 amino acid residues without the signal peptide sequence) that catalyzes the oxidation of coelenterazine to emit light (λ_{max} = 488 nm) and does not require any cofactors in the luminescence reaction [6, 7]. The codon-optimized gene for GLase was expressed in mammalian cells and showed a high luminescence activity when expressed in the ER-Golgi pathway of mammalian cells [3, 7]. The amino-terminal fusion proteins of GLase to prepro-forms of insulin [4] and MMP-2 (Fig. 1c) [5] were correctly processed and secreted from cells. The EM-CCD camera used in our imaging system has high quantum efficiency and the luminescence video images can be recorded at a full spatial resolution of 32 fps (in a model of 512 × 512 pixels). The video-rate bioluminescence imaging using GLase combined with the EM-CCD camera is currently the most suitable method for visualizing protein secretion from a single cell.

1.2 Advantages of Bioluminescence Imaging for Protein Secretion

By detecting luminescence signals from GLase of fusion protein secreted from the outside of the cell surface, time-dependent changes of the amount of secreted protein can be estimated. Bioluminescence imaging is also advantageous for visualization of protein secretion over the whole surface of a cell (Fig. 1b), whereas the fluorescent imaging methods such as total internal reflection fluorescence and two-photon excitation fluorescence have distinct advantages for investigating the dynamics of exocytotic vesicles at limited areas of the cell surface. Furthermore, the bioluminescence imaging method allows long-term imaging of protein secretion in a nontoxic manner, in contrast to the video-rate fluorescence imaging methods that use continuous irradiation with harmful excitation light. As a result, we demonstrated the oscillatory change in the amount of glucose-induced insulin secretion from individual cells over 30 min [4]. Bioluminescence imaging on the whole cell surface facilitates study of the distribution of protein secretion, and we previously demonstrated that the pro-form of MMP-2 was secreted from the basal side of the cell [5].

The cumulative images of maximum luminescence intensity obtained from the video-rate bioluminescence imaging of protein secretion can be used to determine the exocytotic sites of protein secretion in an individual cell. The frequency of exocytotic events can be estimated by counting the luminescence spots of protein secretion in a cell area or by calculating time-dependent changes of maximum luminescence intensity in a micro domain of hot spots for protein secretion. The time-dependent changes of the amount of secreted protein can be estimated by calculating the average luminescence intensity in a single cell and in a region on a cell surface. Thus, the number of exocytotic sites, frequency of exocytotic events, and the amount of secreted protein can be estimated by video-rate bioluminescence imaging.

1.3 Imaging of Both Secreted and Cell Surface-Bound Proteins

MMPs are enzymes that degrade extracellular matrix proteins and regulate cell adhesion and migration. MMP-2 is one of the enzymes that degrades collagen in the basement membrane and plays a major role in cancer cell invasion. MMP-2 is a cell surface-binding protein secreted as the inactive pro-form. On the cell surface, the inactive pro-form of MMP-2 binds to the tissue inhibitor of metalloproteinase-2 (TIMP-2), which associates with the membrane type 1-matrix metalloproteinase (MT1-MMP, also called MMP-14). The amino-terminal peptide of pro-MMP-2 is cleaved by MT1-MMP to give the intermediate form. The intermediate form binds to integrin $\alpha v\beta 3$ at the cell surface, and the fully active MMP-2 is then produced. The polarized localization of MMP-2 on both

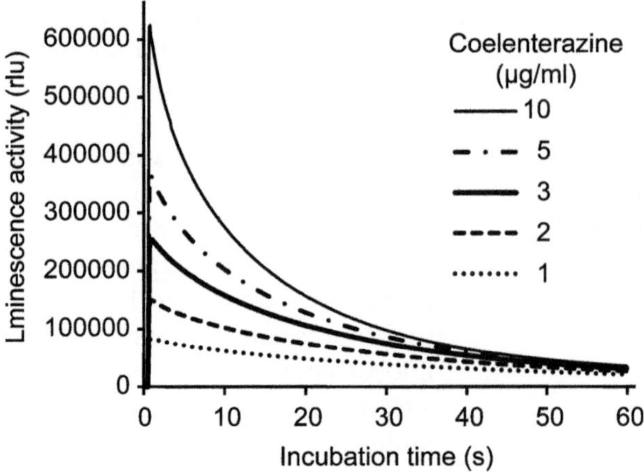

Fig. 2 Luminescence patterns of MMP2-GLase with various concentrations of coelenterazine determined by a luminometer. Luminescence intensity of MMP2-GLase with coelenterazine is recorded for 60 s. HeLa cells were transfected with pcDNA3-hMMP2-GLuc and cultured for 24 h. After removing the medium, the cells were incubated with HBSS for 60 min at 37 °C, and the luminescence activity in HBSS(+) was determined using a luminometer, as described in **Note 9**

lamellipodia and invadopodia of a cell has been shown, and activated MMP-2 is localized at the front of migrating cells with protease activity.

In the video-rate bioluminescence imaging of MMP2-GLase expressed in HeLa cells, two distinct types of secreted MMP-2 and membrane-associated MMP-2 can be visualized on the cell surface in real-time [5]. The exocytotic secretion of MMP2-GLase is observed as transient luminescence spots on the basal side of most cells but at the leading edge of migrating cells [5]. The transient luminescence spots of secreted MMP2-GLase during exocytosis disappear in a few seconds [5]. In addition, MMP2-GLase bound to the basal side of cells can be visualized as the continuous luminescence spots after the addition of coelenterazine [5]. These spots of the membrane-associated MMP2-GLase gradually disappear, corresponding with the time-dependent decrease of the luminescence intensity of GLase (Fig. 2) [5, 8]. The analyses of bioluminescence video images of MMP2-GLase showed repeated secretion of MMP-2 from specific sites at the leading edge of the cell, suggesting that the sites of MMP-2 secretion differ from the MMP-2 binding sites (Fig. 3) [5].

The method of bioluminescence imaging has also been applied for visualization of the transmembrane proteins on the surface of living cells [3, 9]. The proteins are transported to the plasma membrane through the exocytotic pathway similar to other secreted proteins. The method of video-rate bioluminescence imaging described here can be used for visualizing both the exocytotic event and the distribution of transmembrane proteins on the surface of living cells.

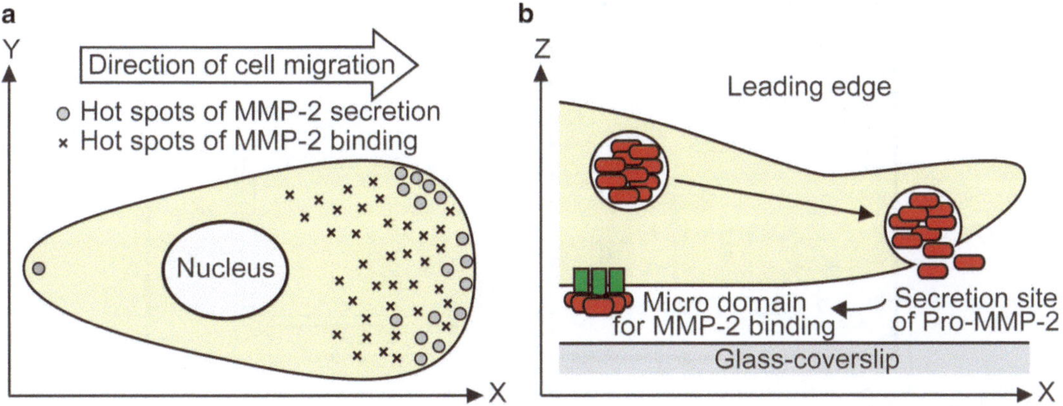

Fig. 3 Schematic representation of MMP-2 secretion and binding on the cell surface in migration cells. (**a**) Hot spots of both secretion and binding of MMP-2 on the cell surface in a migrating cell. (**b**) The inactive pro-form of MMP-2 is frequently secreted on the leading edge and then binds to microregions on the cell surface. Reproduced from Suzuki et al. [5] with the author's copyright under *PLoS One* editorial policies

2 Materials

A schematic representation of our video-rate bioluminescence imaging system is shown in Fig. 4. An inverted microscope equipped with an EM-CCD camera is set in a lightproof box. When a motorized microscope is used, a camera lens adaptor with an infrared (IR)-cut filter should be attached between the microscope and the EM-CCD camera to cut off the light from the pilot lamp in the microscope.

2.1 Bioluminescence Imaging System

1. Microscope: IX71, IX81-ZDC, or IX81-ZDC2 electric motorized inverted microscope (Olympus, Tokyo, Japan).
2. Objective lens: High numerical aperture (NA) oil-immersion (O) or water-immersion (W) objective lens (Olympus): UPLSAPO 20× O/NA 0.95; UPLFLN 40× O/NA 1.30; UPLSAPO 60× O/NA 1.35; UPLSAPO 100× O/NA 1.40; UAPO 20× W3/340/NA 0.70; UAPO 40× W3/340/NA 1.15; and UPLSAPO 60× W/NA 1.20 (*see* **Note 1**).
3. EM-CCD camera: Water-cooled, back-thinned EM-CCD camera (model C9100-13; 512 × 512 pixels, pixel size = 16 μm or model C9100-14; 1,024 × 1,024 pixels, pixel size = 13 μm, Hamamatsu photonics, K.K., Hamamastu, Japan).
4. Camera-mount adaptor with an IR-cut filter (Olympus), which blocks infrared light over 700 nm.
5. Home-made lightproof box.
6. Halogen lamp (LG-PS2; Olympus) as a light source for bright-field microscopic observation (*see* **Note 2**).
7. Thermostat incubator.

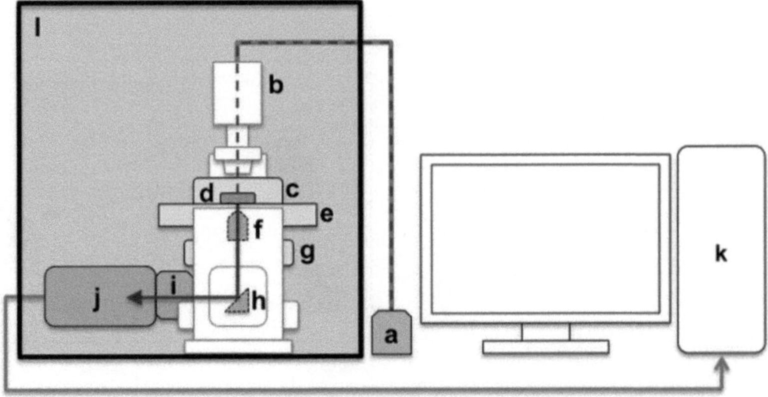

Fig. 4 Video-rate bioluminescence imaging system equipped with EM-CCD camera. A schematic representation of the video-rate bioluminescence imaging system is shown. The inverted microscope equipped with EM-CCD camera is placed in a lightproof box. The representative system consists of halogen lamp (a), inverted microscope (b), thermostat incubator (c), glass-bottom dish (d), stage (e), objective lens (f), ZDC unit (g), prism (h), C-mount adaptor (with an IR-cut filter for the electric motorized microscope) (i), EM-CCD camera (j), the system control computer (k), and lightproof box (l). *Dashed and solid arrows* indicate the light pathway from halogen lamp (a) to EM-CCD camera (j)

8. Electric motorized XY-stage (Ludl Electronic Products Ltd., NY).
9. System control computer (AQUACOSMOS version 2.6 with hard disk recording software; Hamamatsu Photonics).
10. Acquisition setting for video-rate bioluminescence imaging in the software: scan mode, fast scanning; binning, 1 × 1; EM gain, 255 (maximum gain); photon-imaging mode, 1 (*see* **Note 3**).
11. Exposure time: 30–500 ms with an interval time of approximately 0.78 ms per image in the fast scanning mode (in the model of C9100-13).

2.2 Cell Culture and Transfection

1. HeLa cells (ATCC: CCL-2).
2. Culture medium: Dulbecco's modified Eagle's medium with high glucose supplemented with 10 % fetal bovine serum.
3. Culture dishes for passage: Falcon 100-mm culture dishes (BD Bioscience, Bedford, MA).
4. 5 % CO_2 incubator.
5. Glass-bottom dishes for imaging: 35-mm dishes with a glass bottom of Ø = 27 mm (Asahi Glass Co., Tokyo, Japan).
6. FuGENE HD transfection reagent (Promega, Madison, WI).
7. Plasmid vector: pcDNA3-hMMP2-GLuc [5], which is prepared from pcDNA3-GLuc (Prolume Ltd., Pinetop, AZ) (Fig. 1c).

2.3 Bioluminescence Imaging

1. Medium for imaging: Hank's balanced salt solution with Ca^{2+} and Mg^{2+} (HBSS(+); Wako Pure Chemicals, Osaka, Japan).
2. Incubator for pre-warming HBSS(+): Aluminum block bath.
3. Coelenterazine (JNC Co., Tokyo, Japan).

3 Methods

In video-rate bioluminescence imaging for exocytotic protein secretion using GLase, the concentration of coelenterazine in the imaging medium and the exposure time for recording luminescence signals are crucial to obtain clear luminescence images. As the initial intensity and the decay-time of luminescence by GLase are dependent on the coelenterazine concentration (Fig. 2), we optimized this parameter to visualize the secretion of proteins fused to GLase. We found that when 3 μg/ml of coelenterazine is used, over 80 % of the initial luminescence intensity of GLase in the medium is retained within a few seconds (Fig. 2). This concentration of coelenterazine is suitable for visualizing the individual exocytotic protein secretion in a few seconds. After incubation with coelenterazine for over 10 min, coelenterazine is gradually imported into the intracellular network of protein secretion [3]. The weak luminescence signals from intracellular GLase can be detected using an EM-CCD camera with an exposure time of 10–30 s after incubating with 3 μg/ml of coelenterazine for several minutes. With shorter exposure times such as 30–500 ms for imaging, the weak luminescence signals in the cell are below the background level of luminescence. Thus, the initial luminescence intensity of GLase during exocytosis can be principally detected. At the concentrations of 3 μg/ml of coelenterazine, the luminescence signals from the exocytotic secretion of the protein fused to GLase are visible at least for 30–60 min after addition of coelenterazine.

A thermostat incubator on the microscope stage is used to maintain the cultured cells at 37 °C. The HBSS(+) buffer solution used for washing the cells and the imaging medium should be prewarmed at 37 °C to minimize temperature changes. Temperature changes by the addition of HBSS(+) containing coelenterazine cause focus drift during image acquisition using the manually focused microscope.

3.1 Cell Preparation

1. Seed HeLa cells (5×10^4 cells) onto glass-bottom dishes.
2. Transfect with pcDNA3-hMMP2-GLuc (2 μg/dish) using FuGENE HD (6 μl per dish) according to the manufacturer's protocol for 24 h prior to imaging.

3.2 Cell Imaging

1. Initialize the imaging system and the software on the computer. Turn on the water-cooling system for the EM-CCD camera for 30–40 min before image acquisition. Maintain the temperature of the CCD camera at −80 °C when the C9100-13 model is used (or at −70 °C when the C9100-14 model is used). Pre-warm the stage top incubator on a microscope and HBSS(+) at 37 °C.
2. Wash the glass-bottom dish three times with 3 ml of HBSS(+) to rinse away the conditioned medium containing serum, phenol red, and secreted GLase.
3. Add 1 ml of HBSS(+) into the dish, and set the dish on the microscope stage.
4. Focus on the cells under the bright field (*see* **Note 4**).
5. Add coelenterazine in pre-warmed 0.5 ml of HBSS(+) (3 μg/ml; *see* **Note 5**) into the dish, and close the lightproof box.
6. To acquire bioluminescence images of MMP2-GLase secretion, close the shutter of transmission light, set the EM gain level at 255 with the photon-imaging mode at 1, and an exposure time to 500 ms at the initial imaging (*see* **Note 6**).
7. Identify the bioluminescent cells and acquire a video image with an exposure time of 30–500 ms using the hard disk recording software plug-in of the AQUACOSMOS software. The luminescence signals from the exocytotic secretion of MMP2-GLase are observed as transient luminescence spots, which appear in a frame and disappear in a few seconds.
8. Acquire the bright-field images of the cells in the imaging field before and after the recording to analyze the localization of luminescence signals on the individual cells.
9. Video-image acquisition of both secreted and cell surface-binding proteins: Before the addition of coelenterazine (3 μg/ml) to HBSS(+), focus on the basal side of the cells under bright-field conditions. Start continuous auto-focusing with the ZDC2 system to maintain the focus position (*see* **Note 7**). Add coelenterazine in HBSS(+), close the lightproof box quickly, and start to acquire the video image within 10–20 s. Continuous luminescence spots of MMP2-GLase binding on the cell surface and transient luminescence spots of MMP2-GLase secretion can be observed. Continuous luminescence spots of membrane-associated MMP2-GLase decrease to the background level of luminescence within 30 s to 3 min after the addition of coelenterazine (Fig. 2) (*see* **Note 8**).
10. Video-image acquisition of protein secretion from the basal to the surface side of cells: Adjust the focus to the exocytotic luminescence spots on the basal side of the cell, and acquire video images for 2–3 min. Subsequently, shift the focus to the surface side of the cell through *z*-axis position and acquire video images.

3.3 Image Analysis

To analyze the bioluminescence video images, we use the AQUACOSMOS software. In addition, we also use the ImageJ software, (http://rsb.info.nih.gov/ij/index.html).

For example, the data of the bioluminescence video images of both secreted and membrane-associated MMP2-GLase are shown in Fig. 5 [5]. Continuous luminescence spots of the membrane-associated MMP2-GLase and transient luminescence spots of MMP2-GLase secretion can be identified by time-dependent changes in luminescence spots (Fig. 5b, c). The transient luminescence spots of MMP2-GLase secretion (0.5–1 μm diameter) that appeared in a frame can be interpreted as a single exocytotic vesicle, compared with the results of fluorescence imaging of MMP-2 fused to a fluorescent protein. After continuous luminescence signals of the membrane-associated MMP2-GLase disappear, luminescence signals from exocytotic secretion of MMP2-GLase are predominantly observed. The localization of protein secretion sites (Fig. 5d, e), frequency of exocytosis (Fig. 5f), and amount of secreted proteins can be estimated by analyzing the luminescence signals in the video images.

1. Determination of MMP-2 localization (either secreted or bound to the cell surface): Superimpose the pseudo-colored bioluminescence video image on the bright-field image of cells by using a bioluminescence video image after the addition of coelenterazine (Fig. 5a).
2. Identification of the luminescence spots of secreted and membrane-associated MMP-2: Draw a region of interest (ROI) containing a luminescence spot of MMP2-GLase and obtain successive images of the ROI (Fig. 5b). To further analyze time-dependent changes in the luminescence spots, calculate the average luminescence intensities of the ROI in the video image using the "Temporal Analysis" command on the AQUACOSMOS software (Fig. 5c). The image data of the transient luminescence spots (spot 3 and 4 in Fig. 5b, c) represent dynamics of MMP-2 secretion during exocytosis.
3. Determination of the secretion sites of MMP-2: Use successive frames of video images after the continuous luminescence signals of membrane-associated MMP2-GLase have disappeared. Superimpose the pseudo-colored bioluminescence video image on the bright-field image to find the exocytotic secretion sites of cells (Fig. 5d). To analyze the distribution of secretion sites, generate an image consisting of maximum luminescence intensities from the successive frames of a video image using the "MaxTrace" method under the "Sequential calculation" menu on the AQUACOSMOS software. Then, superimpose the pseudo-colored image of maximum luminescence intensity on the bright-field image of cells (Fig. 5e).
4. Determination of the frequency of exocytosis: Draw an ROI containing the secretion sites in the image of maximum

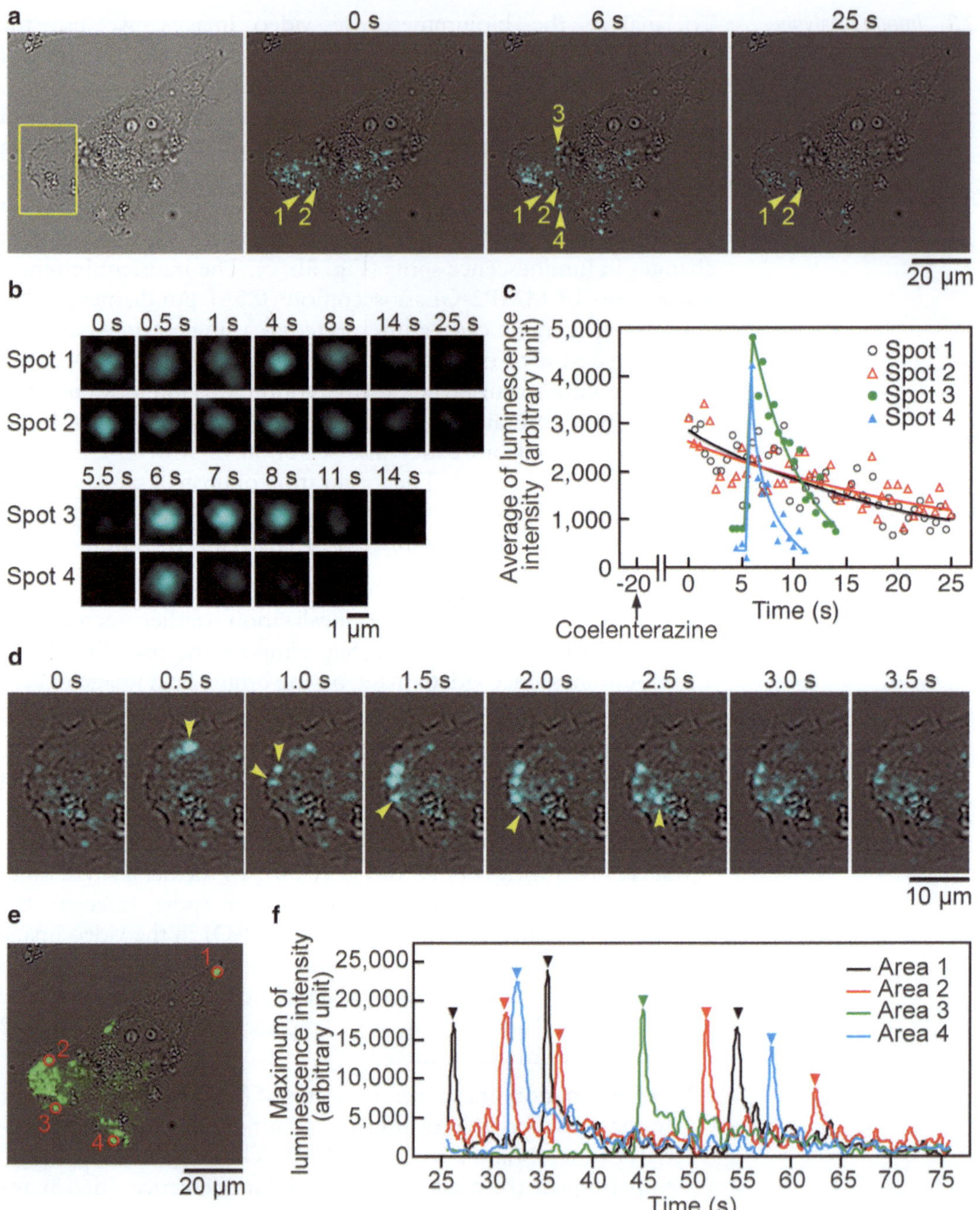

Fig. 5 Video-rate bioluminescence imaging of MMP-2 binding and secretion. Bioluminescence imaging of HeLa cells transiently expressed MMP2-GLase. Objective lens, 40×. Luminescence signals were recorded after 20 s from the addition of coelenterazine (3 μg/ml) in HBSS (+) with an exposure time of 500 ms for 75 s. (**a**) Bright-field image (*left panel*) and luminescence images acquired at 0, 6, and 25 s. Luminescence signals are *cyan-colored* and superimposed on the bright-field image. Two continuous luminescence spots (spot 1 and 2) are indicated by *yellow arrowheads* labeled as 1 and 2, respectively, and two transient diffusive luminescence spots (spot 3 and 4) are indicated by *yellow arrowheads* labeled as 3 and 4, respectively.

luminescence intensity, copy and paste the ROI onto the original video image, and calculate maximum luminescence intensities in the video image using the "Temporal Analysis" command (Fig. 5f). We can also estimate the frequency of exocytosis in the ROI containing several protein secretion sites or a single cell by counting the number of transient luminescence spots of exocytotic protein secretion.

5. Determination of time-dependent changes in the amount of MMP-2 secretion: Draw an ROI containing the MMP-2 secretion sites or a single cell in a bioluminescence video image, and calculate average luminescence intensities in the video image using the "Temporal Analysis" command.

4 Notes

1. To detect weak luminescence signals, the ability of the objective lens to collect light is critical to obtain clear bioluminescence images. The light gathering power is directly proportional to the square of the NA value and indirectly proportional to the square of the magnification value.
2. In our imaging system, the halogen lamp as light source for the microscope is placed outside of the lightproof box and connected to the microscope with a glass fiber cable (Fig. 4). This makes it convenient for changing quickly between the bright-field mode and the bioluminescence-imaging mode. When the halogen lamp is connected to the microscope directly, the lamp should be turned off during the bioluminescence-imaging mode.
3. The photon-imaging mode "1" on the AQUACOSMOS software is critical for enhancing the bioluminescence signals at a full spatial resolution.

(b) Magnified luminescence images of spot 1–4 indicated by *arrowheads* labeled as 1–4 in **(a)**. **(c)** Time-dependent changes of the average luminescence intensities of spots 1–4 indicated by *arrowheads* labeled with 1–4 in **(a)**. **(d)** Magnified luminescence images of MMP2-GLase secretion along the leading edge corresponding to the area indicated by a *yellow square* in **(a)**. Newly appeared luminescence spots are indicated by *yellow arrowheads*. The first image ($T=0$) represents one frame at 51 s 79 ms after the start of recording. **(e)** An image of the maximum luminescence intensity obtained with an exposure time of 500 ms for 50 s after luminescence signals of the membrane-associated MMP2-GLase disappeared. The image was acquired from the beginning at the frame at 26 s 541 ms to the ending at 1 min 16 s 118 ms (100 frames) after the start of recording. Luminescence signals of maximum intensity are *green-colored* and superimposed on the bright-field image. **(f)** Time-dependent changes of the maximum luminescence intensity in the area of *red circles* (1–4) in **(e)**. Reproduced in part from Suzuki et al. [5] with the author's copyright under *PLoS One* editorial policies

4. The bright-field observation with the EM-CCD camera should be performed under the EM-CCD mode at a low EM gain level. After the switch from normal CCD mode to EM-CCD mode, the background signals in the EM-CCD mode at the maximum EM gain 255 in photon-imaging mode "1" are affected during bioluminescence imaging.
5. Keep the stock solution of 1 mg/ml of coelenterazine (dissolved in ethanol) on ice until use. Add the stock solution of coelenterazine to pre-warmed HBSS(+), and further incubate at 37 °C for a few minutes.
6. We use the "custom function keys" for pre-setting bioluminescence and the bright-field observations in the AQUACOSMOS software. When a microscope equipped with mirror units for fluorescence imaging is used, choose the empty position of mirror units in bioluminescence imaging.
7. The bright-field images and the offset position of ZDC in the previous imaging condition of MMP2-GLase secretion are helpful for adjusting the focus on the basal side of cells. Alternatively, co-expression of MMP-2 fused to a fluorescence protein such as mCherry and the fluorescence observation prior to the bioluminescence observation may facilitate focusing. In the cases of a manually focused microscope, the temperature of pre-warmed HBSS(+) containing coelenterazine during addition to the cells should be more strictly controlled to avoid focus drift caused by temperature change.
8. Imaging time for visualization of membrane-associated MMP2-GLase is dependent on the expression level of MMP2-GLase in a cell, exposure time per image in the video imaging, and the concentration of coelenterazine in HBSS(+). For prolonged imaging of MMP2-GLase on the cell surface, we use 1 μg/ml of coelenterazine, because the time-dependent decay of luminescence intensity of GLase is slow as it exhibits the glow-type luminescence pattern (Fig. 2).
9. HeLa cells (2×10^5 cells) cultured in a Falcon 6-well plate (BD Biosciences, Bedford, MA) were transfected with 2 μg of pcDNA3-hMMP2-GLuc and 6 μl of FuGENE HD (Roche Applied Science) and cultured for 24 h. After washing HeLa cells three times with 3 ml of PBS at room temperature, the cells were incubated for 60 min at 37 °C with 1 ml of HBSS(+). Five microliters of the conditioned medium was used for determining the luminescence activity of GLase by adding 50 μl of HBSS(+) containing various concentrations of coelenterazine (JNC Co., Tokyo, Japan). The light intensity was measured using an Atto (Tokyo, Japan) AB2200 luminometer (ver. 2.56, rev. 5.14n) equipped with a R4220P photomultiplier (Hamamatsu Photonics, K.K.).

Acknowledgement

This work was supported in part by a Grant-in-Aid for Scientific Research (C) from the Japan Society for the Promotion of Science and by Strategic Research AGU-Platform Formation (2008–2012) to TS.

References

1. Inouye S, Ohmiya Y, Toya Y, Tsuji FI (1992) Imaging of luciferase secretion from transformed Chinese hamster ovary cells. Proc Natl Acad Sci USA 89:9584–9587
2. Thompson EM, Adenot P, Tsuji FI, Renard JP (1995) Real time imaging of transcriptional activity in live mouse preimplantation embryos using a secreted luciferase. Proc Natl Acad Sci USA 92:1317–1321
3. Suzuki T, Usuda S, Ichinose H, Inouye S (2007) Real-time bioluminescence imaging of a protein secretory pathway in living mammalian cells using *Gaussia* luciferase. FEBS Lett 581:4551–4556
4. Suzuki T, Kondo C, Kanamori T, Inouye S (2011) Video rate bioluminescence imaging of secretory proteins in living cells: localization, secretory frequency, and quantification. Anal Biochem 415:182–189
5. Suzuki T, Kondo C, Kanamori T, Inouye S (2011) Video-rate bioluminescence imaging of matrix metalloproteinase-2 secreted from a migrating cell. PLoS One 6:e25243
6. Tannous BA, Kim DE, Fernandez et al (2005) Codon-optimized *Gaussia* luciferase cDNA for mammalian gene expression in culture and in vivo. Mol Ther 11:435–443
7. Inouye S, Sahara Y (2008) Identification of two catalytic domains in a luciferase secreted by the copepod *Gaussia* princeps. Biochem Biophys Res Commun 365:96–101
8. Verhaegen M, Christopoulos TK (2002) Recombinant *Gaussia* luciferase. Overexpression, purification, and analytical application of a bioluminescent reporter for DNA hybridization. Anal Chem 74:4378–4385
9. Miesenböck G, Rothman JE (1997) Patterns of synaptic activity in neural networks recorded by light emission from synaptolucins. Proc Natl Acad Sci USA 94: 3402–3407

Chapter 7

Bioluminescence Reporter Gene-Based Detection of MicroRNAs

Hae Young Ko, Young Sik Lee, and Soonhag Kim

Abstract

MicroRNAs (miRNAs) are an abundant class of small noncoding RNA molecules that inhibit the expression of cognate genes in multicellular organisms. These small RNAs have been demonstrated to play crucial roles in a variety of biological processes including cell differentiation, proliferation, and survival. Knowledge of specific expression patterns of miRNAs is critical for functional studies. Here, we describe a bioluminescence reporter gene-based method to measure miRNA activity in cultured cells and mice using a *Gaussia* luciferase reporter gene controlled by miRNA binding sites in its 3′untranslated region. This method can be used to noninvasively monitor the expression patterns of functionally active miRNAs involved in different biological processes or diseases in mice.

Key words Bioluminescence reporter gene, *Gaussia* luciferase, MicroRNA, Imaging

1 Introduction

MicroRNAs (miRNAs) are single-stranded, noncoding RNAs of ~22 nucleotides that bind to mRNAs with partial complementarity and induce translational repression and/or mRNA degradation, playing important roles in a wide range of cellular processes [1]. Most miRNA binding sites reside in the 3′ untranslated region (3′UTR) of target mRNAs to mediate posttranscriptional gene regulation [2]. To fully understand the biological functions of miRNAs, it is necessary to determine their spatiotemporal expression patterns. Conventional methods for monitoring the expression of miRNAs include Northern blotting, in situ hybridization, real-time PCR, microarray, and deep sequencing [3]. However, these methods require invasive procedures such as cell disruption and tissue biopsy and thus lack the ability for repetitive detection on the same subjects and in vivo use. More importantly, none of the detection methods do inform whether the expressed miRNA is active for gene silencing. For these reasons, methods to

Christian E. Badr (ed.), *Bioluminescent Imaging: Methods and Protocols*, Methods in Molecular Biology, vol. 1098, DOI 10.1007/978-1-62703-718-1_7,

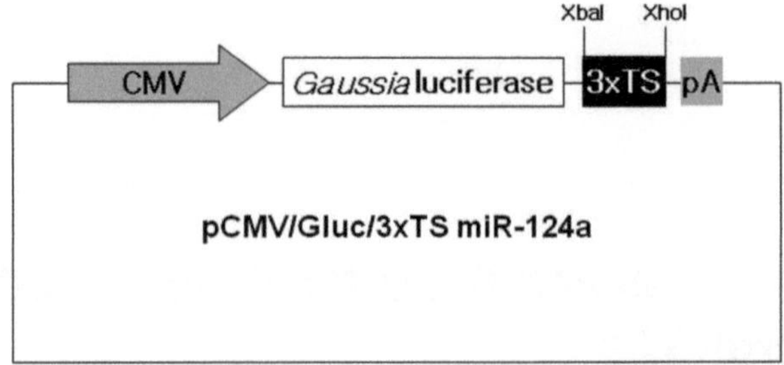

Fig. 1 The schematic diagram of a pCMV/Gluc/3×TS miR-124a construct. The Gluc reporter gene contains three copies of a perfect target site for mature miR-124a (3×TS miR-124a) in the 3'UTR. The expression of the reporter is repressed as miR-124a is targeted to 3×TS of the Gluc mRNA

noninvasively visualize both the expression and activity of a given miRNA are desirable.

Using an alternative approach first described in *Drosophila melanogaster* [4], we previously developed a *Gaussia* luciferase (Gluc) reporter gene-based miRNA imaging system that was designed to detect the expression of functionally active miRNAs in cultured cells and mice. Gluc is much more sensitive than commonly used luciferases from firefly or *Renilla* and catalyzes the oxidation of the substrate coelenterazine in a reaction that emits light with a peak wavelength of 480 nm [5]. The reporter gene has the *Gluc* coding sequence fused to a 3′UTR bearing perfectly complementary binding sites for the miRNA, and thus its expression is controlled by miRNA-directed cleavage of the *Gluc* mRNA through the RNA interference (RNAi) pathway [6, 7] (Fig. 1). Accordingly, an increase in the activity of the target miRNA is reflected by a decrease in Gluc activity. The *Gluc* reporter is constitutively expressed under control of the cytomegalovirus (CMV) promoter in cells lacking the target miRNA. In contrast, where present and active, an miRNA represses the expression of the reporter containing its perfectly complementary binding sites by RNAi in a quantitative manner. We have successfully used this approach to detect the activity of distinct miRNAs in vitro and in vivo whose expression is modulated during differentiation or in tumor cells [8–12]. For example, up-regulation of miR-9, miR-23a, and miR-124a during neuronal differentiation of a mouse embryonic carcinoma cell line (P19 cells) resulted in a decrease in the bioluminescent signal derived from the corresponding miRNA-specific Gluc reporters, indicating that the Gluc activity depends on the levels of functionally active miRNAs [8, 9, 11].

In this chapter, we exemplify bioluminescence reporter gene-based detection of miR-124a activity in culture and in vivo using P19 cells in which miR-124a is gradually up-regulated during neuronal differentiation [11, 13]. We describe detailed protocols for construction of a miR-124a-specific Gluc reporter plasmid and for bioluminescence imaging of miR-124a activity in cell lysates and in nude mice implanted with Gluc-expressing cells.

2 Materials

Prepare all solutions using ultrapure water (18 MΩ dH_2O) and analytical grade reagents. Store all reagents at −20 °C, unless otherwise specified. Follow institutional regulations for waste disposal and animal experiments.

2.1 Recombinant DNA Technique Components

1. *Gaussia* luciferase-expressing vector, pCMV GLUC-1 (Targeting Systems, El Cajon, CA, USA) (*see* **Note 1**).
2. Firefly luciferase-expressing vector (pCMV/Fluc), pGL3-Basic (Promega, Madison, WI, USA) (*see* **Note 2**).
3. 3xTS miR-124a (three tandem copies of a perfect target site for mature miR-124a, which are separated by a spacer of five nucleotides in between) sense and antisense oligonucleotides that once annealed have *Xba*I- and *Xho*I-compatible ends (*see* **Note 3**): 100 μM sense and antisense 3xTS miR-124a oligonucleotides each dissolved in nuclease-free water. 3xTS miR-124a sense (5′-TCGAGaatctagt**TGGCATTCACCGCGTGCCTTAA***tagta***TGGCATTCACCGCGTGCCTTAA***tagta***TGGCATTCACCGCGTGCCTTAAT**-3′; bold, miR-124a target site; underlined, *Xba*I 5′-protruding end and *Xho*I 3′-recessive end; italic lowercase, spacer sequence). 3xTS miR-124a antisense (5′-CTAGA**TTAAGGCACGCGGTGAATGCCA***tacta***TTAAGGCACGCGGTGAATGCCA***tacta***TTAAGGCACGCGGTGAATGCCA**actagattC-3′; bold, miR-124a target site; underlined, *Xho*I 5′-protruding end and *Xba*I 3′-recessive end; italic lowercase, spacer sequence).
4. Annealing buffer (10×): 100 mM Tris–HCl, pH 7.5, 1 M NaCl, and 10 mM EDTA.
5. Restriction enzymes *Xba*I and *Xho*I.
6. Alkaline phosphatase, shrimp.
7. T4 DNA ligase.
8. *Escherichia coli* DH5α-T1^{R} competent cells (Invitrogen) (*see* **Note 4**). Store at −80 °C.
9. LB broth: 0.01 % Bacto-tryptone, 0.005 % Bacto-yeast extract, and 0.01 % NaCl (*see* **Note 5**). Store at 4 °C.
10. LB agar plates containing ampicillin: 0.01 % Bacto-tryptone, 0.005 % Bacto-yeast extract, 0.01 % NaCl, 1.5 % Bacto-agar, and 100 μg/mL ampicillin (*see* **Note 5**). Store at 4 °C.
11. Electrophoresis-grade agarose. Store at room temperature.

12. Agarose gel electrophoresis equipment.
13. Ethidium bromide (EtBr) solution (*see* **Note 6**). Store at 4 °C.
14. Gel-loading buffer (10×): 0.25 % bromophenol blue, 0.25 % xylene cyanol FF, and 50 % glycerol. Store at 4 °C.
15. DNA molecular weight markers.
16. TAE buffer for agarose gel electrophoresis (50×): Store at room temperature.
17. UV transilluminator.
18. T7 primer: 5′-TAATACGACTCACTATAGGG-3′.
19. CMV/reverse 2 primer: 5′-CTTACTTGGCATGACAGTAA-3′.
20. *Taq* DNA polymerase.
21. dNTPs: A mixture of dATP, dCTP, dGTP, and dTTP, each at 100 nM.
22. PCR machine.
23. Incubator and orbital shaker for bacterial culture.
24. NucleoBond Xtra Midi kit (Macherey-Nagel, Düren, Germany). Store at room temperature.

2.2 Culture and Transfection of P19 Cells, and In Vitro Luciferase Assay Components

1. P19 cells (a mouse embryonic carcinoma cell line, ATCC no. CRL-1825).
2. 150 mm tissue culture dishes and 6-well plates.
3. Gelatin.
4. A laminar flow cell culture hood and 37 °C incubator with humidified atmosphere of 5 % CO_2.
5. Trypan blue: 0.5 % in phosphate-buffered saline (PBS).
6. Hemocytometer.
7. Inverted microscope.
8. Growth medium: Minimal essential medium alpha (α-MEM) supplemented with 7.5 % bovine calf serum, 2.5 % fetal bovine serum, and 1 % Antibiotic–Antimycotic solution. Store at 4 °C, which is stable for <3 months.
9. Lipofectamine and Plus reagent (Invitrogen). Store at 4 °C.
10. Opti-MEM medium (GIBCO, Grand Island, NY, USA). Store at 4 °C.
11. Dulbecco's phosphate-buffered saline. Store at 4 °C.
12. Neuronal differentiation medium: Dulbecco's Modified Eagle's medium (DMEM)/F12 [1:1] supplemented with 1 % insulin–transferrin–selenium, 1 % Antibiotic–Antimycotic solution, and 0.5 μM all-*trans*-retinoic acid (RA) (*see* **Note 7**). Store at 4 °C, which is stable for <3 months.
13. Tropix Lysis Solution (Applied Biosystems, Carlsbad, CA, USA). Store at 4 °C.

14. BCA Protein Assay kit.
15. GAR1 buffer and Gaussia substrate (100×) (Targeting Systems, El Cajon, CA, USA)
16. Wallac 1420 VICTOR^3 V multilabel reader (PerkinElmer, Waltham, MA, USA).

2.3 In Vivo Bioluminescence miRNA Imaging Components

1. BALB/c nude mice.
2. PBS. Store at 4 °C.
3. RA: 0.5 μM in absolute ethanol.
4. Insulin syringes: 50 cc, 31 gauge × 8 mm needle.
5. In vivo bioluminescence imaging system (IVIS 100; Caliper Life Science, Hopkinton, MA, USA).
6. In some cases, reagents and equipments described in Subheading 2.2 are also required.

3 Methods

3.1 Generation of a Gluc Reporter Construct for miR-124a

1. Anneal 20 pmol of 3xTS miR-124a sense oligonucleotides with 20 pmol of 3xTS miR-124a antisense oligonucleotides in 1× annealing buffer by heating at 95 °C for 4 min and then cooling slowly to room temperature for approximately 1 h. The annealed oligonucleotide duplexes are flanked by *Xba*I and *Xho*I restriction sites and can be stored at −20 °C.
2. Digest 1 μg of pCMV GLUC-1 vector in a 20-μL reaction containing 1× restriction enzyme buffer with 0.5 μL (10 U) each of *Xba*I and *Xho*I restriction enzymes by incubating the reaction mixture at 37 °C for 2–3 h (*see* **Note 8**). Stop the reaction by inactivating the restriction enzymes at 65 °C for 20 min.
3. To prevent vector self-ligation, dephosphorylate the restricted vector with 1 μL (1 U) of shrimp alkaline phosphatase in a 20-μL reaction containing 1× dephosphorylation buffer for 20 min at 37 °C.
4. Ligate the annealed 3xTS miR-124a oligonucleotide fragment into the digested pCMV GLUC-1 vector (insert:vector molar ratio, 1:1, 1:3, or 3:1) in a 10-μL reaction containing 1× ligase buffer and 1 μL (1 U) of T4 DNA ligase enzyme for 2–3 h at room temperature.
5. Transform an aliquot (1×10^7 cells/200 μL) of competent DH5α-T1^R cells with 3 μL of the ligation mixture from **step 4** according to the manufacturer's instructions (*see* **Note 9**). Incubate transformed bacteria on an LB agar plate containing 100 μg/mL ampicillin for 16 h at 37 °C (*see* **Note 10**).

6. Screen for the recombinant pCMV GLUC-1 plasmid (hereafter referred to as pCMV/Gluc/3xTS miR-124a) bearing 3xTS miR-124a inserted into *Xba*I and *Xho*I restriction sites in the 3′UTR of the encoded *Gluc* reporter gene by colony PCR. For multiple samples (at least ten colonies), make a master mix and aliquot 20 μL in each PCR tube on ice. The 20-μL PCR reaction contains 13 μL sterile water, 2 μL of 10× PCR buffer, 2 μL of 100 nM dNTPs, 1 μL of 10 μM T7 primer, 1 μL of 10 μM CMV/reverse 2 primer, and 1 μL (5 U) of *Taq* DNA polymerase. Add a small amount of a single colony to each reaction mixture by picking the colony with a sterile yellow tip attached to a pipette and pipetting up and down to mix. Perform the PCR reactions using the following cycling parameters: 94 °C for 3 min, followed by 35 cycles of 94 °C for 10 s, 37 °C for 30 s, and 72 °C for 30 s. After PCR, analyze the resulting products by agarose gel electrophoresis to identify a band of 800 bp corresponding to the DNA fragment with 3xTS miR-124a. Pick up one of the candidate colonies from the replica plate and set up a bacterial culture at 37 °C for midiprep of plasmid DNA. Purify the recombinant pCMV/Gluc/3xTS miR-124a plasmid following the manufacturer's protocol and verify it by restriction mapping and DNA sequencing analysis.

3.2 Transfection of P19 Cells and In Vitro Luciferase Assay for miR-124a

We recommend that each experiment includes the following samples in triplicate: (1) Undifferentiated P19 cells, (2) P19 cells undergoing RA-induced neuronal differentiation, (3) Negative control (*Gaussia* luciferase-expressing vector, pCMV GLUC-1 lacking the miR-124a binding site), (4) pCMV/Gluc/3xTS miR-124a (a recombinant pCMV GLUC-1 plasmid bearing three copies of the miR-124a binding site in the 3′UTR). Therefore, prepare reagents sufficient for triplicate samples.

1. Grow P19 cells at 37 °C in the growth medium in a 5 % CO_2-humidified incubator (*see* **Note 11**).
2. One day before transfection, seed 1.5×10^5 cells into each well of 6-well plates and incubate at 37 °C with 5 % CO_2 (*see* **Note 12**).
3. Prepare tube 1 with 1 μg of either pCMV/Gluc/3xTS miR-124a or the empty parental vector pCMV GLUC-1 (negative control) and 4 μL of the Plus reagent in 100 μL of Opti-MEM medium, and incubate for 15 min at room temperature. Next, prepare tube 2 with Lipofectamine (4 μL) and 96 μL of Opti-MEM medium. Add the content of tube 1 to tube 2. Mix by pipetting gently up and down and incubate for 15 min at room temperature.
4. In the meantime, wash the cells twice with Dulbecco's phosphate-buffered saline (2 mL/well is recommended) and add 500 μL of Opti-MEM medium.

5. Add 200 μL of the transfection medium from **step 3** to cells in each well and incubate for 3 h at 37 °C in a 5 % CO_2 incubator.
6. Replace the transfection media with 2.0 mL of the neuronal differentiation medium (for P19 cells undergoing RA-induced neuronal differentiation) or growth medium (for undifferentiated P19 cells), and incubate for an additional 24 h at 37 °C in a 5 % CO_2 incubator.
7. Remove the media from the wells and wash the cells once with Dulbecco's phosphate-buffered saline. Add 100 μL of Tropix Lysis solution to cells in each well and incubate for 15 min at room temperature. Transfer 50 μL of the lysates into each well of a white 96-well plate containing 50 μL of 1× Gaussia luciferase substrate. In addition, prepare a reagent blank solution by mixing 50 μL of Tropix Lysis solution and 50 μL of 1× Gaussia luciferase substrate. Measure luminescence with a Wallac 1420 VICTOR^3 V multilabel plate reader (*see* **Note 13**). Data are presented as the means ± SD of results from triplicate wells.

3.3 In Vivo Bioluminescence Imaging of miR-124a

We recommend that each experiment includes the following samples in triplicate: (1) Undifferentiated P19 cells, (2) P19 cells differentiated by RA treatment, (3) Cell viability control (Firefly luciferase-expressing vector, pCMV/Fluc), and (4) pCMV/Gluc/3xTS miR-124a (a recombinant pCMV GLUC-1 plasmid bearing three copies of the miR-124a binding site in the 3′UTR). Therefore, prepare reagents and mice sufficient for triplicate samples.

1. One day before transfection, plate 2×10^6 cells in 150 mm tissue culture dishes and incubate at 37 °C in a 5 % CO_2 incubator (*see* **Note 12**).
2. To transiently co-transfect P19 cells with pCMV/Gluc/3xTS miR-124a and pCMV/Fluc (cell viability control) (*see* **Note 2**), dilute 5 μg of each plasmid and 60 μL of the Plus reagent with Opti-MEM medium to 500 μL in tube 1 and incubate for 15 min at room temperature. Next, in tube 2, dilute 40 μL of Lipofectamine with Opti-MEM medium to 500 μL. Add the content of tube 1 to tube 2. Mix by pipetting gently up and down and incubate for 15 min at room temperature.
3. In the meantime, wash the cells twice with Dulbecco's phosphate-buffered saline (20 mL/dish is recommended) and add 5 mL of Opti-MEM medium. Add 1 mL of the transfection medium from **step 2** to cells in a culture dish and incubate for 3 h at 37 °C with 5 % CO_2.
4. Replace the transfection media with 25 mL of the growth medium and incubate for an additional 24 h at 37 °C in a 5 % CO_2 incubator.

5. Split the cells into two groups, one to be treated with RA and the other to remain untreated, by resuspending the cells in PBS to a concentration of at least 2.5×10^6 cells/100 μL for each group.
6. To induce neuronal differentiation in one of the two groups of P19 cells, immediately treat the cells with 0.5 μM RA (*see* **Note 14**).
7. Implant untreated cells from **step 5** subcutaneously into the left thigh as a control and RA-treated cells from **step 6** into the right thigh of three 6-week-old BALB/c nude mice, weighing 25–27 g. To do this, pinch the skin of the mouse between index finger and thumb and pull the skin away from the body of the mouse. Inject the cells slowly and evenly into the pouch created by the fingers using insulin syringes with 8 mm × 31 gauge needles for 10 s (*see* **Note 15**).
8. Dissolve 5 mg coelenterazine in methanol up to 1 mL and dilute 20 microliter of the stock solution with PBS to 200 microliter. To acquire Gluc images, inject 5 microgram/50 microliter of coelenterazine via tail vein of each mouse from **step 7** using insulin syringes (*see* **Note 16**).
9. Prior to imaging, anesthetize the mice with 2.5 % isoflurane/O_2 at a flow rate of 1 L/min and administer them through the nose cone attached to a manifold in the imaging chamber (*see* **Note 17**). Sedation normally takes 2–3 min.
10. After anesthesia of the mice, acquire Gluc bioluminescence images using the IVIS 100 equipped with a CCD camera. The imaging time was 5 min after administering the substrate and imaging parameters were as follows: image acquisition time, 1 min; binning, 2; no filter; f/stop (aperture size), 1 (*see* **Note 18**).
11. Three hours after acquiring Gluc images, inject the mice from **step 7** with 3 mg/100 μL (dissolved in PBS) of D-luciferin intraperitoneally to obtain Fluc images (*see* **Note 19**).
12. To acquire Fluc images, repeat **steps 9** and **10** (*see* **Notes 16–18**).
13. Repeat **steps 8–12** for acquisition of time course bioluminescence imaging (Fig. 2).

4 Notes

1. pCMV GLUC-1 vector contains the coding sequence for *Gaussia* luciferase (Gluc) under control of the CMV promoter. A DNA fragment bearing three copies of a perfect binding sequence for a mature miRNA of interest will be placed into *Xba*I and *Xho*I restriction sites located within the 3′UTR between the Gluc coding sequence and the synthetic polyadenylation signal.

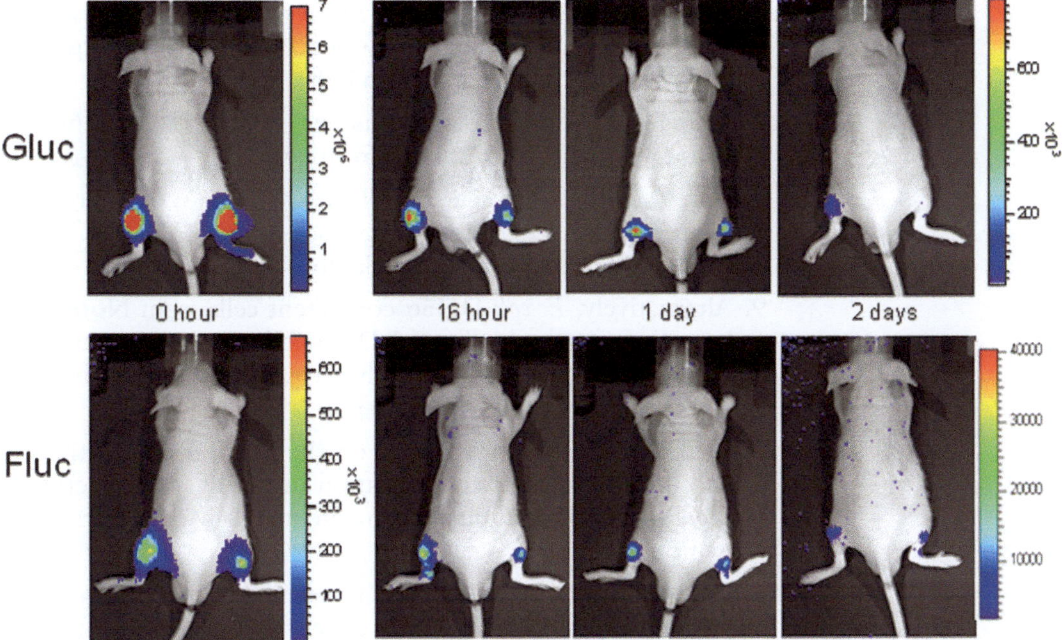

Fig. 2 Bioluminescence images for detecting the expression of mature miR-124a in P19 cells implanted into a nude mouse for 2 days. P19 cells (2.5×10^6) co-transfected with pCMV/Gluc/3xTS miR-124a and pCMV/Fluc were subcutaneously implanted into *left* and *right* thighs of the mouse. For the cells implanted into the *right* thigh, neuronal differentiation of the cells was induced by RA treatment prior to implantation. Gluc bioluminescence in the *right* thigh with RA-treated cells expressing mature miR-124a diminished more rapidly over time than in the *left* thigh. On the other hand, Fluc bioluminescence remained comparable in both thighs over time

2. The miRNA reporter described here is a negative system, i.e., lower expression of the *Gluc* reporter gene in cells with higher activity of a given miRNA. It is necessary to distinguish whether the signal off or reduction attributes to miRNA activity or cell death, particularly for in vivo bioluminescence imaging of miRNAs expressed in Gluc-expressing cells implanted into nude mice. Therefore, cells must be co-transfected with a recombinant pCMV GLUC-1 vector harboring the miRNA binding sites and pCMV/Fluc vector encoding firefly luciferase, which is used to ensure cell viability after implantation.
3. The nucleotide sequence of mature miRNAs can be obtained from the miRNA registry miRBase (http://www.mirbase.org/).
4. Alternatively, the widely used *E. coli* DH5α cells can be made competent by one of the conventional methods available [14].
5. LB medium should be sterilized by autoclaving at 15 psi, 121 °C for 25 min. To prepare LB agar plates containing 100 μg/mL ampicillin, add ampicillin after the sterilized LB agar medium has cooled to 50 °C or below. Pour a thin layer of the medium into petri dishes, let solidify, and store upside down at 4 °C.

6. EtBr is a powerful carcinogen. Always wear gloves, glasses, and lab coat when handling solutions or gels containing EtBr. Protect the EtBr solution from light.
7. RA is light-sensitive and thus the neuronal differentiation medium should be protected from light.
8. Run a standard 1 % agarose gel with an aliquot of the reaction mixture to ensure complete digestions of pCMV GLUC-1 vector (~5.7 kb) as described by Sambrook and Russell [14].
9. Alternatively, *E. coli* DH5α competent cells from **Note 3** can be transformed with ligated DNA following the transformation procedure described by the protocol used for competent cell preparation in **Note 3** [14].
10. If there are very few colonies on the plate, optimize the vector/insert ratio and concentration in the ligation reaction or use commercially available competent cells with high transformation efficiency.
11. Coat tissue culture dishes and plates with sterile 0.1 % gelatin solution to hold P19 cells more securely on the bottom of the cultureware.
12. To achieve the appropriate plating density of viable cells, count the cells using trypan blue and a hemocytometer. P19 cells should be cultured in monolayers for RA-induced neuronal differentiation [15].
13. If low luciferase activities are detected, optimize transfection conditions for the P19 cell line currently used in your lab by adjusting parameters for transfection.
14. Local injection of RA into one thigh can diffuse to the other thigh in the mouse. It is therefore recommended to induce neuronal differentiation of P19 cells by RA in vitro prior to implantation into nude mice.
15. When injecting cells, there should be no air bubbles in the syringe and no leakage from the injection sites.
16. The poor quality of substrate can cause low levels of bioluminescence in vivo. In that case, try to use freshly made substrates (coelenterazine or D-luciferin) and avoid repeated thawing and freezing cycles. Tap the tube containing substrates gently to avoid substrate precipitation immediately prior to injection.
17. Do not inhale isoflurane vapor.
18. In vivo bioluminescence imaging was performed basically according to the manufacturer's instructions. The bioluminescence kinetics should be determined to achieve the imaging time frame for peak bioluminescence of respective luciferase (Gluc or Fluc), when performing baseline imaging.

19. As Gluc activities can interfere with Fluc activities in the same mice, in vivo Fluc activities should be acquired after confirming that the Gluc activity was certainly fade out.

Acknowledgement

This work was supported by the Bio & Medical Technology Development Program of the National Research Foundation (NRF) funded by the Korean government (MEST) (No. 2012-0006097) and a grant of the Korea Healthcare technology R&D Project, Ministry of Health and Welfare (A120254) to SK. This work was also supported by a grant from the National Research Foundation of Korea funded by MEST (No. 2011-0016863) to Y.S.L.

References

1. Wienholds E, Plasterk RH (2005) MicroRNA function in animal development. FEBS Lett 579:5911–5922
2. Bartel DP (2009) MicroRNAs: target recognition and regulatory functions. Cell 136: 215–233
3. Bernardo BC, Charchar FJ, Lin RC, McMullen JR (2012) A microRNA guide for clinicians and basic scientists: background and experimental techniques. Heart Lung Circ 21: 131–142
4. Brennecke J, Hipfinder DR, Strark A et al (2003) Bantan encodes a developmentally regulated microRNA that controls cell proliferation and regulates the proapoptotic gene hid in Drosophila. Cell 113:25–36
5. Tannous BA, Kim DE, Fernandez JL et al (2005) Codon-optimized Gaussia luciferase cDNA for mammalian gene expression in culture and in vivo. Mol Ther 11:435–443
6. Hutvágner G, Zamore PD (2002) A microRNA in a multiple-turnover RNAi enzyme complex. Science 297:2056–2060
7. Zeng Y, Wagner EJ, Cullen BR (2002) Both natural and designed micro RNAs can inhibit the expression of cognate mRNAs when expressed in human cells. Mol Cell 9: 1327–1333
8. Lee JY, Kim S, Hwang do W et al (2008) Development of a dual-luciferase reporter system for in vivo visualization of MicroRNA biogenesis and posttranscriptional regulation. J Nucl Med 49:285–294
9. Ko MH, Kim S, Hwang do W et al (2008) Bioimaging of the unbalanced expression of microRNA9 and microRNA9* during the neuronal differentiation of P19 cells. FEBS J 275:2605–2616
10. Kim HJ, Chung JK, Hwang DW et al (2009) In vivo imaging of miR-221 biogenesis in papillary thyroid carcinoma. Mol Imaging Biol 11: 71–78
11. Ko HY, Hwang do W, Lee DS et al (2009) A reporter gene imaging system for monitoring microRNA biogenesis. Nat Protoc 4: 1663–1669
12. Kang WJ, Cho YL, Chae JR et al (2012) Dual optical biosensors for imaging microRNA-1 during myogenesis. Biomaterials 33: 6430–6437
13. Huang B, Li W, Zhao B et al (2009) MicroRNA expression profiling during neural differentiation of mouse embryonic carcinoma P19 cells. Acta Biochim Biophys Sin 41:231–236
14. Joseph S, David WR (2001) Molecular cloning: a laboratory manual. Cold Spring Habor, New York
15. Pacherník J, Bryja V, Esner M et al (2005) Neural differentiation of pluripotent mouse embryonal carcinoma cells by retinoic acid: inhibitory effect of serum. Physiol Res 54: 115–122

Chapter 8

Monitoring of Transcriptional Dynamics of HIF and NFκB Activities

Miguel A.S. Cavadas and Alex Cheong

Abstract

Genetic experiments over the last few decades have identified many regulatory proteins critical for DNA transcription. The dynamics of their transcriptional activities shape the differential expression of the genes they control. Here we describe a simple method, based on the secreted luciferase, to measure the activities of two transcription factors NFκB and HIF. This technique can effectively monitor dynamics of transcriptional events in a population of cells and be up-scaled for high-throughput screening and promoter analysis, making it ideal for data-demanding applications such as mathematical modelling.

Key words NFκB, HIF, Transcription, Luciferase, Mathematical modelling

1 Introduction

The quantification of promoter activity using luciferase reporter is a common molecular biology technique based on the bioluminescent output of the enzymatic reaction. The commonly used reporters are the firefly [1] and the renilla [2] luciferases, and such an experiment would require lysis of the cultured cells at a given time point in order to obtain a measure of promoter activity. This limits the amount of data which can be obtained from such an experiment. Recently, a new generation of luciferase reporters such as Gaussia and Cypridina has been reported [3, 4] with the unique characteristic of being secreted out of the cell. As an improvement over the original first generation luciferases, the secreted Gaussia or Cypridina enzymes can be collected from the culture media, thus enabling repeated measurements of promoter activity from the same population of cells (Fig. 1).

Two important transcription factors are activated during hypoxic inflammation, resulting in distinct and interconnected signalling networks: the hypoxia-inducible factor HIF [5] and the pro-inflammatory factor NFκB [6]. We have started to apply this new reporter assay in increasing our understanding of the transcriptional

Christian E. Badr (ed.), *Bioluminescent Imaging: Methods and Protocols*, Methods in Molecular Biology, vol. 1098,
DOI 10.1007/978-1-62703-718-1_8,

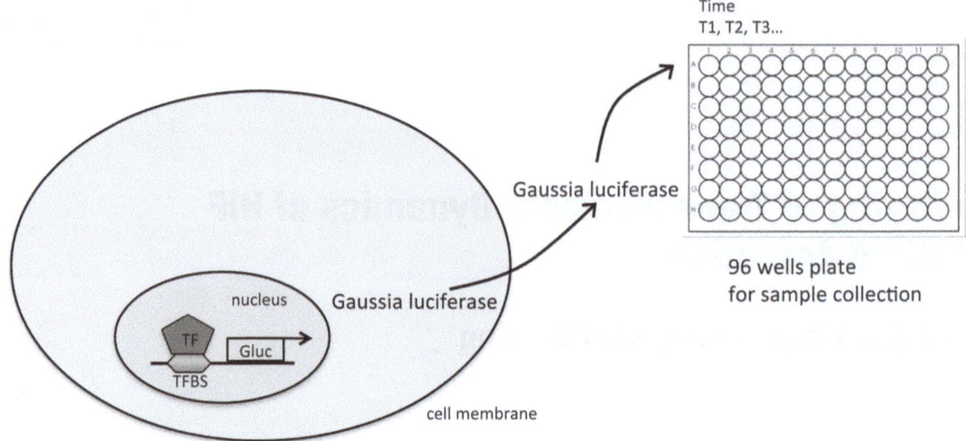

Fig. 1 Gaussia luciferase assay. The cells are transfected with a Gaussia vector under the control of a desired transcription factor (TF). Upon binding to its response element (TFBS) on the vector, Gaussia luciferase is produced and secreted out of the cell into the culture media. The media is collected at desired time interval (T1, T2, T3) into a 96-well plate. Gaussia substrate is later added to the plate and bioluminescence measured

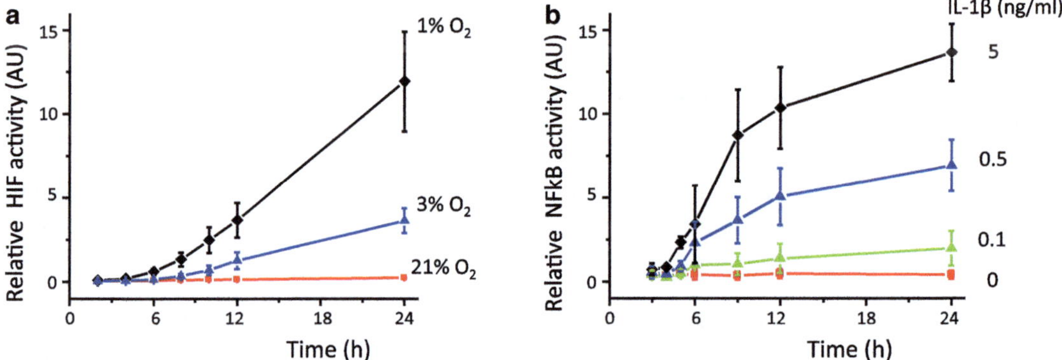

Fig. 2 Examples of HIF or NFκB activity measured using Gaussia luciferase reporter. Transcriptional dynamics of (**a**) the hypoxia-inducible factor (HIF) in varying oxygen tension in cells transfected with a Gaussia luciferase reporter under the control of HIF and (**b**) NFκB in a range of interleukin-1b (IL-1β) in cells transfected with a Gaussia luciferase reporter under the control of NFκB. Data shown is adapted from Bruning et al. [7] and is the mean of four experiments, and error bars represent the standard error of mean

dynamics of the HIF and NFκB upon stimulation [7–9]. Thus we could explore a dynamic range of stimulation which translates to promoter activity for both HIF (oxygen) and NFκB (IL-1β) (Fig. 2) and investigate synergy between the two transcription factors [7]. Furthermore, the assay can be up-scaled to test several stimulations at the same time, generating a “high-throughput” pipeline which we have used for mathematical modelling of the HIF signalling network [9].

Coupled with other molecular biology techniques such as real-time RT-PCR and chromatin immunoprecipitation, the secreted luciferase reporter assay can provide solutions to gain insights into temporal transcriptional dynamics.

2 Materials

2.1 Reporter Constructs

1. pGluc-Mp vector (made from pGluc-basic vector (NEB) with customized minimal promoter inserted (*see* **Note 1**)).
2. pTK-Gluc vector (NEB).
3. pCMV-Cluc vector (NEB).
4. Synthesized DNA sequence containing binding sites for NFκB with restriction recognition sites for *Bgl*II and *Eco*RI (*see* **Note 2**).
5. Synthesized DNA sequence containing binding sites for HIF with restriction recognition sites for *Bgl*II and *Eco*RI (*see* **Note 3**).
6. Cloned DNA sequence of the human cyclooxygenase-2 gene with restriction recognition sites for *Bgl*II and *Eco*RI (*see* **Note 4**).
7. Restriction enzymes *Bgl*II and *Eco*RI.
8. Agarose.
9. Gel purification kit (Quiagen).
10. T4 DNA ligase.
11. *E. coli* competent cells for subcloning.
12. Luria broth agar.
13. Luria broth.
14. Plasmid DNA extraction kit (Promega).
15. Spectrophotometer.

2.2 Cell Culture and Transfection

1. Human embryonic kidney cells (HEK293) (*see* **Note 5**).
2. Dulbecco's Modified Eagle's Medium (DMEM) including 10 % fetal calf serum and 1 % penicillin and streptomycin.
3. Optimem media (Invitrogen).
4. Lipofectamine 2000 (Invitrogen).

2.3 Measuring Gaussia Luciferase

1. 96-well white plates (Nunc).
2. Plate reader able to read luminescence.
3. Gaussia luciferase substrate (coelenterazine).

3 Methods

3.1 Making the Reporter Constructs

1. Digest the DNA sequences and the pGluc-Mp vector with the restriction enzymes *Bgl*II and *Eco*RI.
2. Run the digested DNA sequences in a 2 % agarose gel, cut and gel purify the correct DNA fragments with the gel purification kit.
3. Ligate the DNA sequences to the opened pGluc-Mp vector using T4 DNA ligase.
4. Transform the ligated DNA into *E. coli.* competent cells.
5. Pick colony from agar plate (*see* **Note 6**).
6. Amplify the bacterial clone in Luria broth overnight.
7. Extract the plasmid DNA with the Plasmid DNA extraction kit.
8. Measure the amount of DNA obtained using a spectrophotometer (*see* **Note 7**).
9. Have the DNA sequenced to verify correct insertion (*see* **Note 8**).

3.2 Cell Transfection

1. Seed HEK293 cells into 24-well culture plates at a density of 100,000 cells per well the day before transfection (*see* **Note 9**).
2. For transfection of 1 well, mix 100 ng of required DNA into 100 μL of Optimem, then add 0.5 μL of Lipofectamine 2000.
3. Leave for 30 min at room temperature before adding to the well.
4. Cells should be transfected and ready to use the day after.

3.3 Activation of HIF

1. For activating HIF, we use a hypoxia chamber where the cells are kept at normal atmospheric pressure with 5 % CO_2, and the desired O_2 and N_2 (e.g., 1 % O_2 and 94 % N_2).
2. The culture media must be preequilibrated in the hypoxia chamber overnight to the desired oxygen tension. At the start of the experiment in the hypoxia chamber, remove culture media and replace with fresh hypoxia-equilibrated media (*see* **Note 10**).
3. Alternatively, HIF can be activated pharmacologically using hypoxia mimetics such as dimethyl-oxaloylglycine (DMOG).
4. Pipette 10 μL of the culture media from each well every 2 h into a 96-well white plate (*see* **Note 11**).

3.4 Activation of NFκB

1. For activating NFκB, we have used interleukin-1beta (IL-1β) at a range of concentration (0–5 ng/ml).
2. At the start of the experiment, remove culture media and replace with fresh media with or without IL-1β.
3. Pipette 10 μL of the culture media from each well every 2 h into a 96-well white plate.

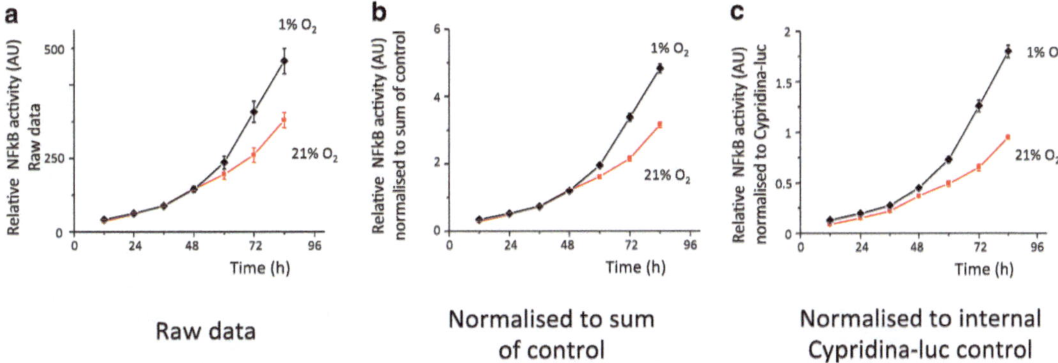

Fig. 3 Normalization of Gaussia luciferase data. HEK293 cells were transfected with a Gaussia luciferase construct under the control of NFκB and cultured either in normoxia (21 % O_2) or hypoxia (1 % O_2). Raw luciferase data was plotted as a function of time (**a**) or first normalized to either (**b**) the sum of experimental values for control normoxia (mathematical normalization) or (**c**) the values of the co-transfected Cypridina luciferase. Data shown is the mean of four experiments, and error bars represent the standard error of mean

3.5 Analyzing Transcriptional Activity

1. Reconstitute the Gaussia luciferase substrate as per manufacturer's recommendation (*see* **Note 12**).
2. Add the substrate to the wells and read the luminescence in a plate reader (*see* **Note 13**).
3. The values obtained from the plate reader can be transferred to a spread sheet (such as Microsoft Excel) for data analysis.
4. In order to compare among experiments, the raw data can be averaged out to produce a mean response (Fig. 3a). However it is preferable to normalize it first. This can be done by different methods:
   (a) Mathematically by dividing by the sum of the values for the control condition (Fig. 3b; *see* **Note 14**).
   (b) By dividing by the total protein in each wells at the end of the experiment (*see* **Note 15**).
   (c) By dividing by a constitutively active luciferase reporter which does not use the same type of substrate as Gaussia luciferase (e.g., firefly or Cypridina) (Fig. 3c; *see* **Note 16**).
5. The data can be represented as either a plot of relative luciferase over time or as fold activity over control (Fig. 4). For the latter, the data needs to undergo either a first-order derivation or a linear regression to obtain the rate of transcription (*see* **Note 17**).

3.6 Limitations of the Gaussia Luciferase

We found that the Gaussia luciferase is stable only within a range of pH. If the pH of the culture media is too acidic (which can occur if the transfected cells are overgrown or when the CO_2 level in the incubator is too high) or too alkaline (when the CO_2 level in the incubator is too low), the activity of the luciferase will be decreased (Fig. 5).

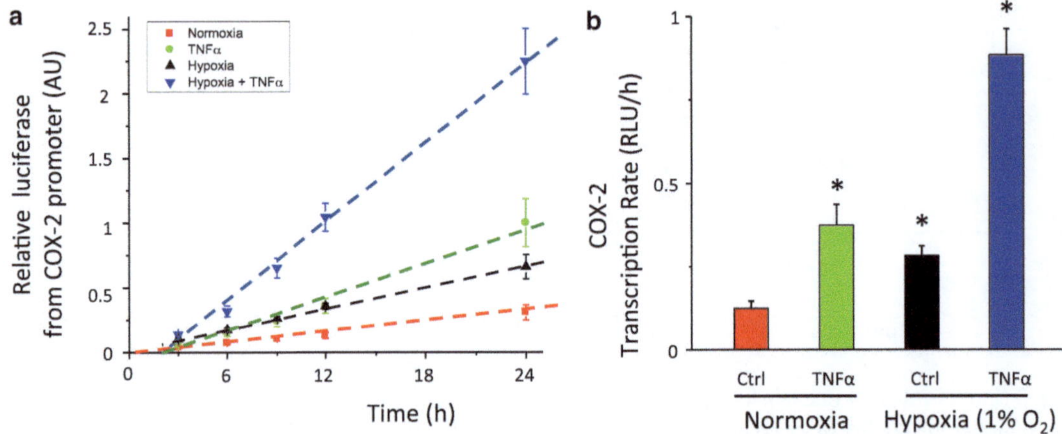

Fig. 4 Analysis of Gaussia luciferase data. HEK293 cells were transfected with a Gaussia vector containing a piece of the human cyclooxygenase-2 promoter (which includes both HIF and NFκB response elements) and stimulated with either hypoxia (1 % O_2), inflammation (1 ng/ml TNFα) or hypoxic inflammation (1 % O_2 and 1 ng/ml TNFα). The resulting luciferase data can be represented either as (**a**) a time-course plot or (**b**) transcriptional rates relative to control unstimulated. Transcriptional rates are obtained by linear regression of the data points (*dotted lines*). Data shown is adapted from Bruning et al. [7] and is the mean of eight experiments, and error bars represent the standard error of mean. Significant difference ($p < 0.05$) is denoted by *asterisks*

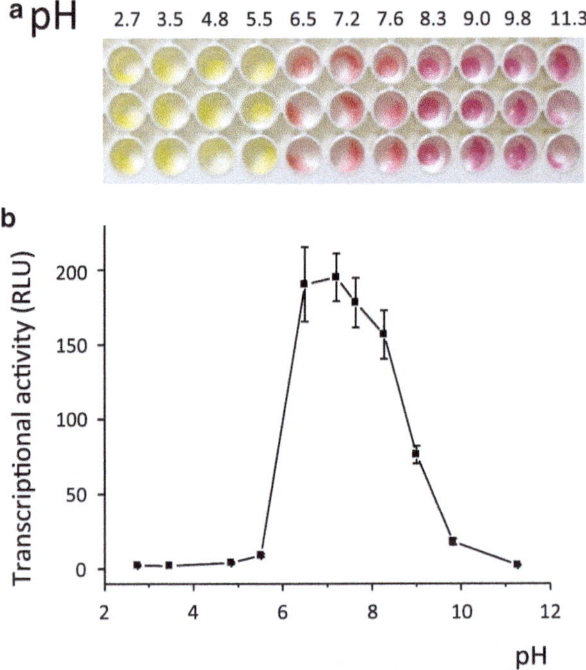

Fig. 5 Effect of pH on the Gaussia luciferase activity. (**a**) Photograph of cell culture media samples from cells transfected with pGluc-HRE. The pH of the media was changed using either sodium hydroxide or hydrochloric acid. (**b**) Plot of Gaussia luciferase activity as a function of pH Data shown is the mean of three experiments, and error bars represent the standard error of mean

4 Notes

1. In this paper, we have customized the basic Gaussia luciferase vector by inserting a minimal promoter to create the pGluc-Mp vector. Alternatively, it is also possible to purchase Gaussia vectors with a range of minimal promoters such as mini-TK (from the Herpes Simplex Virus thymidine kinase promoter).
2. We have here used a tandem repeat of NFκB binding sites to maximize the promoter activity.
3. This sequence corresponds to four copies of the promoter sequence of the erythropoietin gene previously shown to bind to HIF [10].
4. We have used the promoter sequence corresponding to the sequence -4 to -631 of the human cyclooxygenase 2 gene, which includes both HIF and NFκB binding sites [7]. This was cloned from human genomic DNA.
5. We have successfully used the Gaussia luciferase reporter assay in a variety of cell types. We here show data from HEK293 cells as an example. Consult the datasheet of the transfection reagents to verify that the desired cells can be efficiently transfected.
6. To rapidly screen a multitude of potential positive clones on the agar plate, we have used part of the bacterial clones as DNA template for RT-PCR. The primers are designed to amplify the region before and after the cloning site. Thus only the DNA from the positive clones will be amplified by the primers.
7. We use the Nanodrop spectrophotometer (Thermo Scientific) as it uses a small volume (1 μL) for measuring DNA concentration.
8. There are many free software for analyzing and archiving DNA sequences. We personally use ApE (A plasmid Editor; http://biologylabs.utah.edu/jorgensen/wayned/ape/).
9. We normally ensure that there are 2 wells for each of the conditions to be measured. The measurements from the 2 wells will be averaged to give a mean value.
10. The volume we use is 400 μL for each well. This gives good coverage of the cells and is not drastically reduced by repeated sampling.
11. We pipette the culture media directly into a 96-well plate and freeze the plate.
12. The substrate is very stable, especially after the inclusion of the stabilizing agent (included in the NEB Gaussia substrate pack), and can be stored in the freezer for later use. Unfortunately, we did not observe the same stability with the Cypridina substrate, which needs to be assembled and used for one measurement.

13. If the samples from the experiment span more than one more plate, there should be an inter-plate control to be included in each plate. The values of the inter-plate control should be similar. If they are not, the values need to be adjusted and the adjustment ratio is also applied to all the measurements from the plate.
14. This is a mathematical method to normalize a set of data [7]. Simply add all the values of the control condition and use it to divide all the values from the experiment. This can only be used when there are no missing numbers (Fig. 3b).
15. Simply lyse the cells at the end of the experiment and perform a protein assay to determine the protein concentration in each well. The protein lysate can also be used for quantifying other proteins by Western blot.
16. The constitutively active luciferase reporter must be transfected at the same time as the Gaussia luciferase construct. The principle is similar to the standard dual luciferase assay which uses firefly and renilla luciferases (Fig. 3c).
17. Gaussia luciferase accumulates in the culture media. Thus the rate of Gaussia accumulating in the media is considered to be proportional to the rate of transcription. This rate is obtained by first-order derivation, i.e., calculating the difference between two data points divided by the time interval between the two points. Linear regression is similar, except that this assumes the transcription rate to be linear over the range of time measured (as shown in the example in Fig. 4). This enables statistical testing of the changes in the transcriptional rates due to a stimulus.

Acknowledgement

This work was supported by Science Foundation Ireland (grant number 06/CE/B1129).

References

1. de Wet JR, Wood KV, DeLuca M et al (1987) Firefly luciferase gene: structure and expression in mammalian cells. Mol Cell Biol 7(2):725–737
2. Lorenz WW, McCann RO, Longiaru M, Cormier MJ (1991) Isolation and expression of a cDNA encoding Renilla reniformis luciferase. Proc Natl Acad Sci U S A 88(10):4438–4442
3. Tannous BA, Kim DE, Fernandez JL et al (2005) Codon-optimized Gaussia luciferase cDNA for mammalian gene expression in culture and in vivo. Mol Ther 11(3):435–443. doi: 10.1016/j.ymthe.2004.10.016
4. Nakajima Y, Kobayashi K, Yamagishi K et al (2004) cDNA cloning and characterization of a secreted luciferase from the luminous Japanese ostracod, Cypridina noctiluca. Biosci Biotechnol Biochem 68(3):565–570
5. Eltzschig HK, Carmeliet P (2011) Hypoxia and inflammation. N Engl J Med 364(7):656–665. doi:10.1056/NEJMra0910283

6. Taylor CT (2008) Interdependent roles for hypoxia inducible factor and nuclear factor-kappaB in hypoxic inflammation. J Physiol 586(Pt 17):4055–4059. doi:10.1113/jphysiol.2008.157669
7. Bruning U, Fitzpatrick SF, Frank T et al (2012) NFkappaB and HIF display synergistic behaviour during hypoxic inflammation. Cell Mol Life Sci 69(8):1319–1329. doi:10.1007/s00018-011-0876-2
8. Bruning U, Cerone L, Neufeld Z, Fitzpatrick SF et al (2011) MicroRNA-155 promotes resolution of hypoxia-inducible factor 1alpha activity during prolonged hypoxia. Mol Cell Biol 31(19):4087–4096. doi:10.1128/MCB.01276-10
9. Nguyen LK, Cavadas MAS, Scholz CC et al (2013) A dynamic model of the hypoxia-inducible factor (HIF) network. J Cell Sci 126(Pt 6):1454–1463
10. Wang GL, Semenza GL (1993) General involvement of hypoxia-inducible factor 1 in transcriptional response to hypoxia. Proc Natl Acad Sci U S A 90(9):4304–4308

Chapter 9

Real-Time Bioluminescent Tracking of Cellular Population Dynamics

Dan Close, Tingting Xu, Steven Ripp, and Gary Sayler

Abstract

Cellular population dynamics are routinely monitored across many diverse fields for a variety of purposes. In general, these dynamics are assayed either through the direct counting of cellular aliquots followed by extrapolation to the total population size, or through the monitoring of signal intensity from any number of externally stimulated reporter proteins. While both viable methods, here we describe a novel technique that allows for the automated, non-destructive tracking of cellular population dynamics in real-time. This method, which relies on the detection of a continuous bioluminescent signal produced through expression of the bacterial luciferase gene cassette, provides a low cost, low time-intensive means for generating additional data compared to alternative methods.

Key words Bacterial luciferase, *lux*, Optical imaging, Cell culture, Population tracking, Screening

1 Introduction

Monitoring the dynamics of a cultured eukaryotic cellular population can be useful for determining the presence of estrogenic [1] or androgenic compounds [2], determining drug efficacy [3], toxicity screening [4], or a host of other [5]. While data generated from this type of experiment is straightforward and highly reliable, the main detraction of this technique has been its inability to provide real time data that can be monitored in an automated fashion [6]. This restriction, however, is not a functional limitation of the assay logistics, but rather is due to the nature of the reporter systems that have been employed to visualize the subject cells. Direct cell counting provides a means to overcome this hurdle, but has proven to be prohibitive due to the large investment of time associated with the calculation of cell density in an aliquot of the total population and the related likelihood of error that is encountered when extrapolating this number to an estimate of total population size. Therefore, the use of optical reporters such as firefly luciferase or GFP remains as the preeminent method for assaying cell population dynamics. However, while these

Christian E. Badr (ed.), *Bioluminescent Imaging: Methods and Protocols*, Methods in Molecular Biology, vol. 1098,
DOI 10.1007/978-1-62703-718-1_9, © Springer Science+Business Media New York 2014

reporter systems allow for the determination of cellular population size with a high level of accuracy, neither are amenable to real-time dynamics tracking. In the case of firefly luciferase, this is because sample destruction is required prior to data acquisition in order to treat the cells with an expensive chemical luciferin that induces bioluminescent production. GFP, on the other hand, emits a fluorescent output signal that does not require the destruction of the cell, but is limited due to the relatively high level of background fluorescence present in eukaryotic cells during the imaging process, which results in a low signal-to-noise ratio and decreases the sensitivity of the assay [7–8].

To circumvent these detractions, cells can be tagged via expression of a modified bacterial luciferase (*lux*) gene cassette. The genes of the *lux* cassette (*luxCDABEfrp*) encode both a luciferase protein and the additional proteins capable of generating all required substrates for its function [9]. Therefore, when the full complement of *lux* genes are expressed simultaneously within a cell, it becomes capable of generating a constitutively active bioluminescent signal that can be continuously monitored across any experimentally relevant timeframe.

2 Materials

2.1 Cell Culture Medium

The exact composition of cell culture medium will be dependent on the cell line being used. In our hands this procedure has been successful regardless of the medium used, so it is therefore recommended that the cell culture medium suggested by the cell manufacturer be used throughout the course of the experiment (*see* **Note 1**).

2.2 Transfection Reagents

1. Opti-MEM Reduced Serum Medium (Life Technologies).
2. Lipofectamine 2000 transfection reagent (Life Technologies) (*see* **Note 2**).
3. Plasmid vector DNA containing the full *lux* gene cassette (*luxCDABEfrp*) in its human-optimized form (*see* **Note 3**).
4. Plasmid vector DNA containing only the *luxA* and *luxB* genes and their associated linker region.

2.3 Reagents for the Selection of Successfully Transfected Cells

1. Light assay reagent (*see* **Note 4**): 4 μM riboflavin 5′-monophosphate sodium salt, 0.2 % (w/v) bovine serum albumin (BSA), 0.002 % (w/v) n-decanal, and 1U NAD(P)H:FMN Oxidoreductase protein from *Photobacterium fischeri* prepared in sterile water.
2. 1 mM *β*-NADPH tetrasodium salt (*see* **Note 4**).
3. 0.05 % Trypsin.

4. 0.0067 M Phosphate Buffered Saline (PBS), pH 6.8.
5. Appropriate selection antibiotic(s) for the vector(s) containing the full complement of *lux* genes.

2.4 Cell Culture Equipment

1. Class II biological safety cabinet.
2. Temperature controlled, CO_2 regulated incubator.

2.5 Imaging Equipment

The expressed *lux* genes generate a bioluminescent signal at a wavelength of 490 nm. Therefore, almost any standard photomultiplier tube (PMT) or charge coupled device (CCD) camera-based imaging equipment will be suitable for observing and recording the resultant bioluminescent signal (*see* **Note 5**). The most important consideration is that the imaging device be capable of supporting the type of plate that is used in the experiment (96-well, 384-well, etc.). For the selection of intermediately transfected cell lines, a single tube luminometer such as the LB 9509 (Berthold Technologies) will facilitate bench top screening, but the procedure can be performed in the same plate reader used for bioluminescent imaging if needed.

3 Methods

3.1 Development of an Intermediately Transfected Cell Line

Because the *lux* gene cassette consists of six separate gene productions, and all six of these are required for bioluminescent production, a two-step transfection process is recommended. The first step will introduce only the *luxA* and *luxB* genes in order to provide an area of homology that significantly improves the efficiency and speed of autobioluminescent cell line development (*see* **Note 6**). All steps should be performed in a class II biological safety cabinet to prevent contamination of cell cultures unless otherwise stated.

1. Perform a kill curve on the untransfected stock of the cell line that will receive the *lux* cassette genes (*see* **Note** 7).
2. The day prior to transfection, trypsinize cells from their flask and obtain a total cell count using a hemocytometer, Coulter counter, or other cell counting system.
3. Resuspend the harvested cells in two wells of a six-well plate (*see* **Note 8**) in a 2 ml volume of the appropriate medium so that they will be ~85 % confluent at the time of transfection (*see* **Note 9**).
4. Allow the cells to incubate at 37 °C and 5 % CO_2 for 24 h.
5. In a 2 ml micro-centrifuge tube, add 10 μl of Lipofectamine 2000 to 140 μl of pre-warmed Opti-MEM medium.
6. Prepare a second 2 ml micro-centrifuge tube by adding 2.5 μg of the vector containing only the *luxA* and *luxB* genes

and bringing the total volume to 150 μl with pre-warmed Opti-MEM medium.

7. Remove the full 150 μl volume from one of the two micro-centrifuge tubes and combine into the remaining tube. Mix by gently flicking the tube.
8. Allow the combined mixture to incubate at room temperature for 5 min.
9. Carefully pipette 250 μl of the combined mixture into one of the two wells of the six-well plate (*see* **Note 10**). The second well will be left unmodified as a negative control.
10. Incubate the cells at 37 °C and 5 % CO_2 for 24 h.

3.2 Selection of the Intermediately Transfected Cell Line

Cells should be kept in an incubator at 37 °C and 5 % CO_2 except during manipulation.

1. Trypsinize the transfected cells to detach them from the six-well plate, then dilute them at a ratio of 1:100 with fresh medium, and replate 2 ml volumes into five of the six wells of a new six-well plate.
2. Repeat this process with the negative control well and replate into the remaining well. Return the cells to the incubator for 24 h.
3. Twenty-four hours post plating, remove the medium from all wells with serological pipettes, then wash the cells once with 1 ml of PBS and refresh with 2 ml of the appropriate medium containing selective antibiotic.
4. Monitor the cells daily and refresh with selective medium every 3rd day until all cells in the negative control well have died and distinct colonies have formed in the remaining five wells.
5. Transfer surviving colonies into individual wells of a 24-well plate in a 1 ml volume of the selective medium and return to the incubator.
6. Upon reaching 80 % confluence, trypsinize the cells from their wells and transfer them to T_{25} flasks for continued growth.
7. When cells in a T_{25} flask reach confluence, passage them at a ratio of 1:5 into a new T_{25} flask, and harvest the remainder of the cells by transferring the spent medium to a 15 ml centrifuge tube and centrifuging at 300 × *g* for 7 min.
8. Remove the supernatant, resuspend the cell pellet in 1 ml of PBS and transfer to a micro-centrifuge tube. Place the new tube immediately on ice.
9. Perform a protein extraction of the cell pellet by repeating three cycles of 30 s submersion in liquid nitrogen, followed by 3 min thawing in a 37 °C water bath.

10. Centrifuge the extracted sample at 14,000×*g* in a bench top microfuge for 10 min and transfer the resulting supernatant to a new micro-centrifuge tube. Remove an aliquot of the extracted protein solution and perform a BCA protein assay to determine the overall concentration of soluble protein in the sample, and then place the tube on ice until it is ready to be processed (*see* **Notes 11** and **12**).
11. In an appropriate cuvette for the luminometer, combine 400 μl of the extracted protein with 500 μl of the light assay reagent.
12. Spike in 100 μl of the NADPH solution and immediately measure the luminescent output using a 1 s integration time. This result should be normalized to the total soluble protein concentration as determined using a standard protein determination assay (for example BCA assay) in order to control for variations in cell number during harvesting.
13. Select the intermediately transfected cell line whose protein extract produced the greatest normalized level of luminescent output for use in developing a stable autobioluminescent cell line. This will ensure that the *luxAB* genes were successfully integrated into the cellular genome and provide homologous recombination insertion points for the full *lux* cassette that are preselected for an improved likelihood of expression.

3.3 Development of the Fully Transfected Autobioluminescent Cell Line

Follow **steps 1–10** of Subheading 3.1, however, using 2.5 μg of the plasmid DNA containing the full complement of the *lux* genes rather than the plasmid DNA containing only the *luxA* and *luxB* genes in **step 6** (*see* **Note 13**).

3.4 Selection of Fully Transfected Autobioluminescent Cell Lines

Cells should be kept in an incubator at 37 °C and 5 % CO_2 except during manipulation.

1. Trypsinize the transfected cells to detach them from the six-well plate, then dilute them at a ratio of 1:100 with fresh medium, and replate 2 ml volumes into five of the six wells of a new six-well plate.
2. Repeat this process with the negative control well and replate into the remaining well. Return the cells to the incubator for 24 h.
3. Twenty-four hours post plating, remove the medium from all wells with serological pipettes, then wash the cells once with 1 ml of PBS and refresh with 2 ml of the appropriate medium containing selective antibiotic.
4. Monitor the cells daily and refresh with selective medium every 3rd day until all cells in the negative control well have died and distinct colonies have formed in the remaining five wells.

5. Transfer surviving colonies into individual wells of a 24-well plate in a 1 ml volume of the selective medium and return to the incubator.
6. Upon reaching 80 % confluence, trypsinize the cells from their wells and transfer them to individual T_{25} flasks for continued growth.
7. When cells in the T_{25} flask reach confluence, passage them at a ratio of 1:5 into a new T_{25} flask. Harvest the remainder of the cells, count them to determine the total cell number, and then transfer them to a 15 ml centrifuge tube and centrifuge at $300 \times g$ for 7 min.
8. Resuspend the cell pellet in 1 ml of the appropriate medium and transfer immediately to an individual well of a 24-well pate (*see* **Note 14**).
9. Assay the cells for autobioluminescent production in a plate reader using a 1 min read time every 30 min for 24 h (Fig. 1) (*see* **Note 15**) and normalize the bioluminescent readings to the total cell number in each well.
10. The cell line displaying the greatest level of normalized autobioluminescent production should then be selected for all future experiments (*see* **Note 16**).

3.5 Imaging Autobioluminescent Cells

1. Allow a T_{25} flask of autobioluminescent cells to grow to 75 % confluence under standard growth conditions.
2. Trypsinize the actively growing cells to detach them from the T_{25} flask and obtain total cell counts using a hemocytometer, Coulter counter, or other cell counting system.
3. Resuspend 5×10^4 autobioluminescent cells into triplicate wells of a 24-well plate (*see* **Note 17**), along with an equal number of untransfected cells (also in triplicate) to act as the negative control.
4. Incubate the cells at 37 °C and 5 % CO_2 for 2 h to allow the cells to attach to the plate (*see* **Note 18**).
5. If required, treat the cells with your compound(s) of interest (*see* **Notes 19** and **20**).
6. Put the plate into the plate reader and measure the autobioluminescence continuously over a 24 h period using a 10 s read time (Fig. 1) (*see* **Notes 21** and **22**).

4 Notes

1. If available, it is advantageous to choose a version of the appropriate medium that does not contain phenol red. It has been observed that the presence of phenol red in the medium can lead to a slight absorption of resulting autobioluminescent signal.

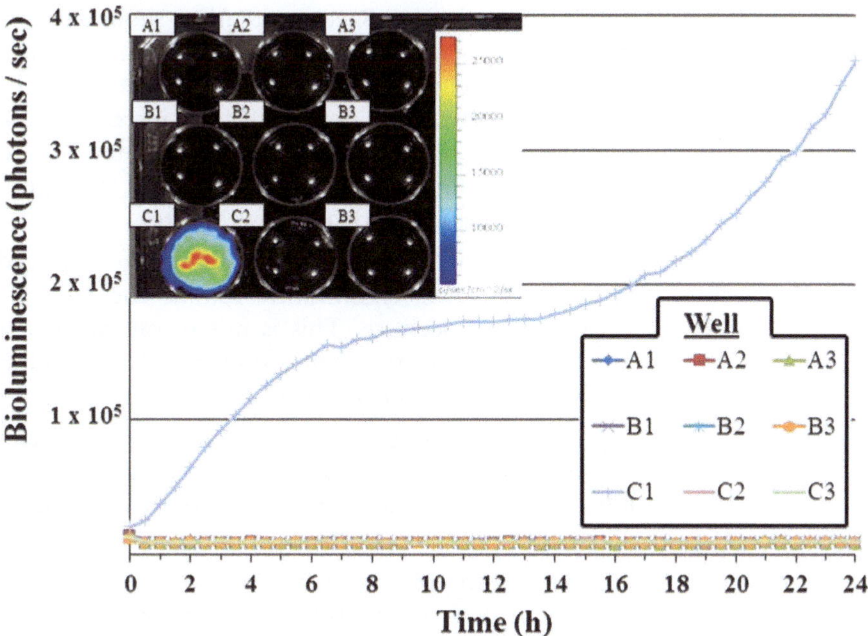

Fig. 1 The *lux* cassette genes allow for continuous bioluminescent production, permitting facile detection of positively transfected stable cell lines and providing an elegant means for tracking population changes over time

However, this reduction in signal strength is very minor, and if phenol red is a necessary component of the appropriate medium its negative absorption effects will be less than those resulting from growth in sub-optimal medium.

2. Alternative transfection reagents based on polycationic liposomes are theoretically viable options for this protocol; however, all of our work to date has utilized Lipofectamine 2000 and this protocol is therefore optimized for use with this reagent. If a different reagent is used, consult the manufacturer to determine necessary changes to the transfection protocol.
3. A human optimized version of the *lux* gene cassette consists of a vector or series of vectors that contains the *luxA*, *luxB*, *luxC*, *luxD*, and *luxE* genes of *Photorhabdus luminescens* and the *frp* gene of *Vibrio harveyi* in their human codon-optimized forms. While these genes do not necessarily need to be housed on a single plasmid to generate a bioluminescent phenotype [10], it is much easier and faster to generate bioluminescent cell lines by using a single vector containing each of the required *lux* genes under the control of a single strong promoter and linked via viral 2A elements.
4. This must be made fresh before each experiment.
5. While any standard plate reader or CCD camera-based detection system can be used, an instrument with integrated incubation capabilities will significantly improve the automated nature of the

imaging experiment. Special attention should be paid to the humidity level that can be maintained within the imaging chamber because an instrument that maintains an elevated temperate (i.e., 37 °C) but does not control for increased humidity will hasten the rate at which the medium evaporates. This is especially true as well volume decreases during higher throughput experimental designs (i.e., with a ≥96-well plate).

6. This step can be skipped (moving straight to Subheading 3.3); however, this will significantly reduce the efficiency of the transfection procedure. This is not recommended unless the goal is to develop transiently transfected autobioluminescent cell lines for short term (16–36 h) experiments.
7. It is very important that a new kill curve be conducted prior to the start of the transfection protocol. The resistance of cell lines to antibiotics can vary greatly both from lab to lab and within the same line as it ages.
8. The size of the plate can be varied as needed; however, the amount of both DNA and Lipofectamine 2000 that are used must then be scaled appropriately. For information on scaling the amount of DNA and Lipofectamine 2000 when alternative size plates are used, consult the manufacturer's manual at the time of purchase.
9. For a standard six-well plate, we have found that plating between 4×10^5 and 6×10^5 cells will provide optimal confluence for transfection at 24 h post plating.
10. Care must be taken not to dislodge the plated cells. It is therefore advisable to gently swirl the medium within the well while adding the mixture solution dropwise. This ensures that the attached cells are not adversely affected, as well as evenly distributes the mixture solution within the well.
11. This extracted protein sample can be stored at −80 °C or in liquid nitrogen indefinitely if needed. It is not recommended that the sample be stored at −20 °C or 4 °C for longer than 1 h, or loss of protein activity can occur that will invalidate the results of the impending light assay. Following transfer to ice, the sample should be processed as quickly as possible to insure the best results.
12. It is not necessary to wait until the results of the BCA assay are known to proceed. The light assay can be performed immediately and, when complete, can be normalized using the resulting BCA data.
13. Note that a second kill curve must be performed on the intermediately transfected (*luxAB* containing) cell line in order to determine its resistance to the antibiotic selection marker that will be used for selection of the stably transfected autobioluminescent cell line.

14. This step can be performed in any size plate as needed; however, the volume of medium should be scaled appropriately to account for the change in well size. In general, it is not recommended that plates with fewer than 24 wells be used, since the higher concentration of cells across a lower area provides for improved signal detection. In addition, an opaque plate (white or black) should be used, as the use of a clear plate can lead to signal contamination during screening.
15. The read time can be adjusted based on the sensitivity of the plate reader; however, it is not recommended that times be shorter than 10 s or longer than 10 min. The 24 h duration may be much longer than is necessary to determine which lines are autobioluminescent, but a gradual increase in bioluminescent production is often observed following transfer to the 24-well plate as the cells adapt to the new environment and this time period can serve to determine which cell lines have stable, rather than transient, autobioluminescent production (Fig. 1).
16. This cell line is now stably transfected and can be stored in liquid nitrogen for future use if needed.
17. The number of cells required is dependent on the size of the well being used. For assays performed in 24-well plates, the minimum number of cells that allow for reliable detection is 5×10^4. However, for smaller wells a lower cell number can be used and detected with the same reading window time (i.e., 1×10^4 cells can be imaged in a 96-well plate). More cells can always be included to increase the autobioluminescent signal if the cell line was not efficiently transfected. If a CCD camera-based instrument is used, a minimum number of 5×10^3 cells should be used for assays performed in 24-well plates.
18. This step is optional if the attachment of cells is integral to the experimental design. Readings can be taken immediately if attachment is not required. The attachment time varies greatly with the cell line employed and may have to be determined empirically due to the effects of gene integration during the transfection reactions.
19. The compound being tested should first be prepared as a concentrated stock solution. The final concentration, which will be based on the requirements of the experimental design, should therefore be determined based on what its concentration will be post-application.
20. If an organic solvent is used to prepare the compound, the volume of compound solution added to the cells should not exceed 0.2 % of the total volume of medium in each well to prevent vehicle-based cytotoxic effects that could potentially mask any compound-mediated effects.

21. The reading window can be adjusted based on the requirements of the experimental design. While imaging for between 10 s and 1 min is common in PMT-based plate readers, if a CCD camera-based instrument is being used to obtain pseudocolor images, a longer integration time of up to 10 min may be used to improve the quality of the visual image.
22. Up to 24 h continuous reading is recommended for testing acute toxicity or other fast-acting effects. If longer term monitoring is necessary (i.e., for determining the proliferative effect of a target compound) the cells can be imaged once every 24 h and should be kept in the incubator between readings.

References

1. Soto A et al (1995) The E-SCREEN assay as a tool to identify estrogens: An update on estrogenic environmental pollutants. Environ Health Perspect 103:113–122
2. Yang X et al (2011) Novel membrane-associated androgen receptor splice variant potentiates proliferative and survival responses in prostate cancer cells. J Biol Chem 286:36152–36160
3. Rudin M, Weissleder R (2003) Molecular imaging in drug discovery and development. Nat Rev Drug Discov 2:123–131
4. Houck K, Kavlock R (2008) Understanding mechanisms of toxicity: Insights from drug discovery research. Toxicol Appl Pharmacol 227:163–178
5. Contag C, Bachmann M (2002) Advances in in vivo bioluminescence imaging of gene expression. Annu Rev Biomed Eng 4: 235–260
6. Close D et al (2011) Light without substrate amendment: the bacterial luciferase gene cassette as a mammalian bioreporter. *Proc SPIE* 8029: 80291F-1-80290F-10
7. Close D et al (2011) Comparison of human optimized bacterial luciferase, firefly luciferase, and green fluorescent protein for continuous imaging of cell culture and animal models. J Biomed Opt 16:047003
8. Tsien R (1998) The green fluorescent protein. Annu Rev Biochem 67:509–544
9. Meighen E (1991) Molecular biology of bacterial bioluminescence. Microbiol Rev 55:123–142
10. Close D et al (2010) Autonomous bioluminescent expression of the bacterial luciferase gene cassette (*lux*) in a mammalian cell line. PLoS ONE 5:e12441

Chapter 10

Fabrication of Bioluminescent Capsules and Live-Cell Imaging

Sung Bae Kim

Abstract

The plasma membrane of living cells is an interface of material transfers and an antenna for outer signals. This chapter provides a guide on how to fabricate bioluminescent capsules for illuminating intracellular signaling and cargo protein delivery. The capsule consists of four components, which are, in consecutive order: a secretion peptide (SP), a host luciferase body (leader), a guest protein or peptide (cargo), and a membrane localization signal (MLS). Any guest protein, including a luciferase or a fluorescent protein, may be sandwiched between the host luciferase body and MLS and may be deliverable to the plasma membrane (PM), where the capsule waits for outer signals and to quickly release the embedded luciferase in response to a specific signal. The present strategy provides an efficient molecular vehicle for cargo proteins and imaging of intracellular molecular events in living cells without substrate-derived demerits of luciferases.

Key words Apoptosis, Bioluminescent capsule, Bioluminescence, Luciferase, Membrane localization, Imaging, Substrate

1 Introduction

Although bioluminescence is an emerging optical readout, it requires excess substrate and molecular oxygen (O_2) as a prerequisite, which hampers its use in living cells and complex organisms [1]. In contrast to recent achievements in optical intensity [2, 3], the substrate-derived problems of luciferases, e.g., the membrane permeability of the substrates and the biased diffusion of the substrate into the intracellular compartments of mammalian cells, remain a major drawback in live-cell imaging [4]. In particular, organelle-sequestered luciferases should immediately consume nearby coelenterazine (CTZ) and O_2 and may fall into substrate anoxia in living mammalian cells or tissues. Under a deficiency of substrates and O_2, the optical intensities should be dominated simply by the rates of the substrate and O_2 supply.

To address the substrate-derived problems, the author took note of the privileges of the PM. The PM defines the cell boundaries

Christian E. Badr (ed.), *Bioluminescent Imaging: Methods and Protocols*, Methods in Molecular Biology, vol. 1098, DOI 10.1007/978-1-62703-718-1_10,

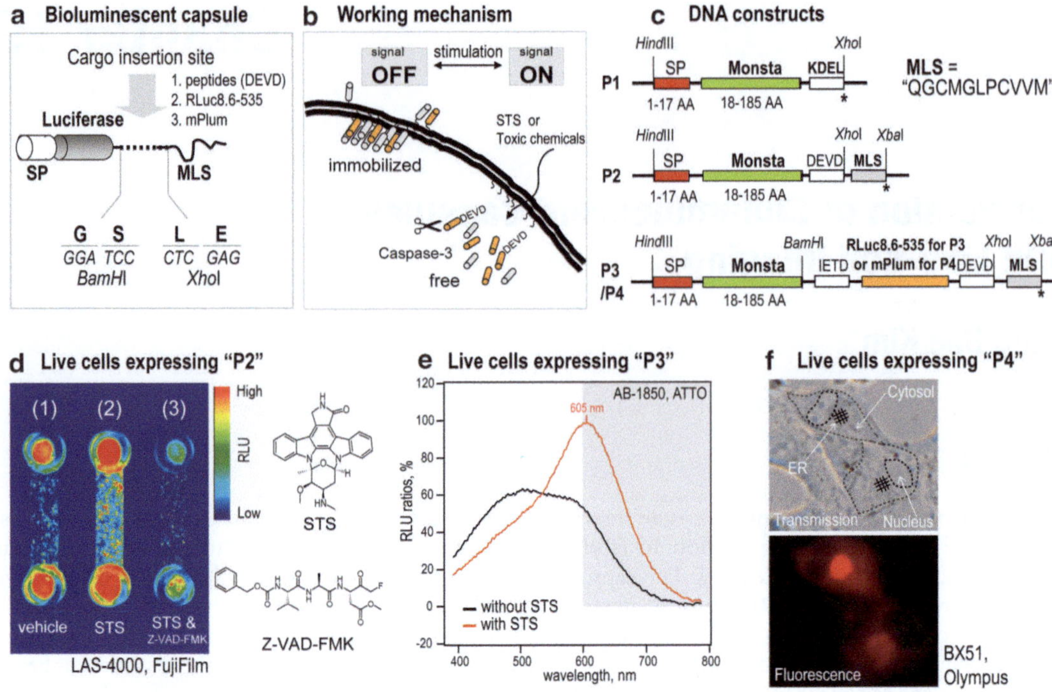

Fig. 1 Live-cell imaging with bioluminescent capsules. (**a**) A bioluminescent capsule. A luciferase leader (host) is located at the head and followed by a guest protein. The capsule ends with the membrane localization signal (MLS). (**b**) The working mechanism of a bioluminescent capsule. In the presence of apoptosis signaling, the hinge regions of the capsules are dissected by intrinsic caspases. (**c**) cDNA constructs encoding bioluminescent capsules. (**d**) Visualization of apoptosis signaling with a bioluminescent capsule in living cells. (**e**) Dramatic convergence of the bioluminescence spectra from living cells carrying a bioluminescent capsule according to apoptosis signals. (**f**) Fluorescence image of live cells carrying a bioluminescent capsule. The plasma membrane (PM) and the endoplasmic reticulum (ER) are illuminated by the capsule. Reproduced from Kim et al. 2012 with permission from Bioconjugate Chemistry (ACS) [4]

and acts as a key interface that senses external signals, allowing the cells to change their behavior in response to environmental cues [5]. The PM has particularly distinct privileges over other cellular ingredients on live-cell imaging; e.g., (1) a bioluminescent indicator on the PM immediately responds even to poor membrane-permeable stimulators, (2) infinite amounts of substrates and O_2 should be supplied to the probe, and (3) bioluminescence imaging (BLI) is easy on the PM because the entire cell territory is bright and the long-term optical intensity on the PM is stable.

The present chapter provides a detailed guide on how to create a bioluminescent capsule located in the PM, where the capsule waits for outer signals and signal-dependently provokes the release of the embedded luciferase. The capsule consists of a secretion peptide (SP), a host luciferase body a guest protein or peptide (cargo), and a membrane localization signal (MLS) (Fig. 1a). Any guest proteins with specific functionalities may be deliverable by the capsule to the endoplasmic reticulum (ER) and PM. We exemplify the visualization of an apoptosis signaling with the capsule, which carries caspase substrate peptides comprising Asp-Glu-Val-Asp (DEVD) or Ile-Glu-Thr-Asp (IETD) sequences (Fig. 1b). Not only the peptides but also the fluorescent proteins and luciferases were found to be efficiently portable by the capsule.

2 Materials

2.1 Construction of Bioluminescent Capsules

1. cDNAs encoding Monsta (custom-synthesized) [6, 7].
2. cDNA encoding *Renilla reniformis* luciferase 8.6-535 (RLuc8.6-535) [8].
3. cDNA encoding fluorescent protein "mPlum" (Clontech).
4. A mammalian expression vector, pcDNA3.1(+) (Invitrogen).
5. Custom-made primers for polymerase chain reaction (PCR).
6. A ligation kit with a ligase, Restriction enzymes (*Hind*III, *BamH*I, *Xho*I), and DNA polymerase.

2.2 Live-Cell Imaging with Bioluminescent Capsules

1. Native coelenterazine (nCTZ) as the substrate of Monsta and RLuc8.6-535 in a luciferase assay kit (Promega; cat. E2820).
2. A CTZ derivative, ViviRen® (Promega).
3. Microslides for cell culture and live-cell imaging (2.5 × 7.5 cm, μ-slide $VI^{0.4}$; cat. 80606; ibidi, Germany).
4. Hanks' balanced salt solution (HBSS) buffer, pH 7.4.
5. TransIT-LT1® (Mirus) as a lipofection reagent.
6. Staurosporine (STS) as an apoptosis inducer.
7. Carbobenzoxy-valyl-alanyl-aspartyl-[*o*-methyl]-fluoromethylketone (Z-VAD-FMK) as an apoptosis inhibitor.
8. Dulbecco's modified Eagle's medium (DMEM) for cell culture, which is supplemented with 10 % fetal bovine serum (FBS) and 1 % of a mixture of penicillin and streptomycin (P/S).
9. African green monkey fibroblast-derived COS-7 cells as a mammalian cell. For all the experiments, culture the COS-7 cells in DMEM supplemented with 10 % FBS and 1 % P/S.
10. Dimethyl sulfoxide (DMSO).
11. A serum-free medium, e.g., Opti-MEM I®.
12. Potassium iodide (KI), an additive for boosting bioluminescence intensity (final concentration: 50 mM).

2.3 Instrumentation

1. GenomeLab GeXP (BeckMan Coulter) as a DNA sequencer.
2. Tpersonal (Biometra) as a thermal cycler.

3. LAS-4000 (Fujifilm) as an image analyzer, equipped with a specific image analysis software (Multi Gauge 3.1; Fujifilm).
4. AB-1850 (ATTO) as a charge-coupled device (CCD) spectrophotometer for accurate light collection in the red region [9].
5. BX51 (Olympus) as a fluorescent microscope, equipped with a specific control software (AxioVision v4.6).

3 Methods

Carry out all the live-cell imaging procedures at room temperature.

3.1 Creation of a Mammalian Expression Vector Encoding a Bioluminescent Capsule

1. Generate cDNA encoding a host luciferase, Monsta (1–185 AA; *see* **Notes 1** and **2**), by polymerase chain reaction (PCR) with corresponding forward and reverse primers (*see* **Note 3**) and templates to introduce unique restriction sites (*Hind*III, *BamH*I, and *Xho*I) and short flexible linkers (*see* **Note 4**), according to the design of DNA constructs shown in Fig. 1c.
2. Extend both ends of cDNAs encoding a guest protein, RLuc8.6-535 or mPlum (*see* **Note 5**), by multiple PCR to introduce specific nucleotides encoding an Asp-Glu-Val-Asp (DEVD) peptide, an Ile-Glu-Thr-Asp (IETD) peptide (*see* **Note 6**), and/or a membrane localization signal (MLS; *see* **Note 7**), as shown in Fig. 1c.
3. Ligate the cDNA fragments with the use of a ligase to make the chimera constructs in Figure (C) and subclone these into a mammalian expression vector, e.g., *p*cDNA 3.1(+) vector carrying a cytomegalovirus (CMV) promoter and ampicillin resistance gene.
4. Confirm the sequences of the subcloned constructs in the plasmids with the use of a DNA sequencer (*see* **Note 8**).
5. Name the generated plasmids in consecutive order as *p*P1, *p*P2, *p*P3, and *p*P4, as shown in Fig. 1c. Name the corresponding fusion proteins consecutively as **P1**, **P2**, **P3**, and **P4** after their expression.

   The above procedure is briefly illustrated in Fig. 2.

3.2 Relative Optical Intensity of the Capsules in Living Cells

1. Grow COS-7 cells in the channels of a microslide (μ-slide $VI^{0.4}$, ibidi) until the cell density reach approximately 85 % confluency in a CO_2 incubator (5 % CO_2, 37 °C) (*see* **Note 9**).
2. Transiently transfect the COS-7 cells with an aliquot of *p*P1 (ER-locating) or *p*P2 (PM-locating) plasmids dissolved in a lipofection reagent (TransIT-LT1), where the mixing ratio of the transfection cocktail is [DNA (plasmid)–TransIT-LT1 (repofection agent)–Opti-MEM (serum-free medium) = 1 μL:3 μL:100 μL]. The cells are then incubated in a CO_2 incubator (5 % CO_2, 37 °C) for 16 h (*see* **Note 10**).

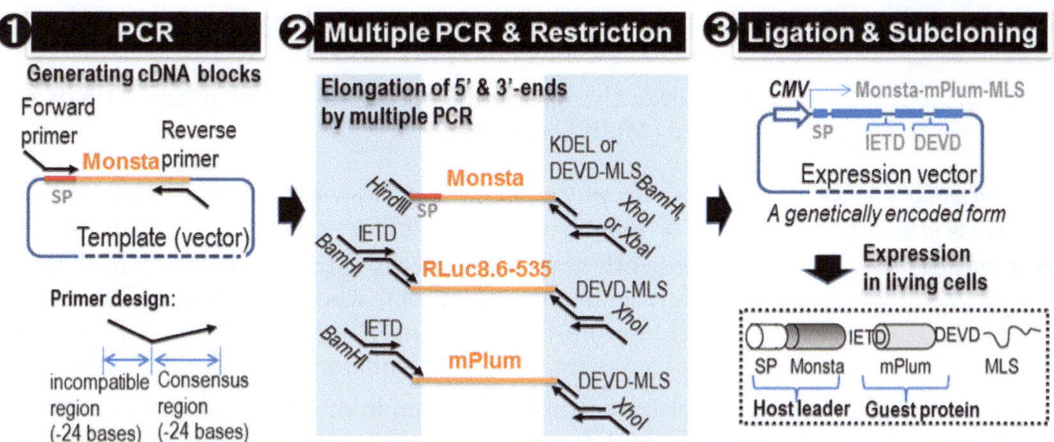

Fig. 2 Generation procedure for genetically encoded bioluminescent capsules. First, cDNA blocks are generated by PCR. The 5′ and 3′ ends of the blocks are further elongated by multiple PCR to create hinge regions, which carry caspase substrate peptides and a membrane localization signal. The blocks are ligated and subcloned into a mammalian expression vector for live-cell imaging

3. Remove the culture medium from the microslide and simultaneously inject 100 μL of native coelenterazine (nCTZ)-dissolved HBSS buffer (nCTZ–HBSS = 1:100) (*see* **Notes 11** and **12**) into the 6 channels of the microslide with the use of a multichannel pipet (*see* **Note 13**).
4. Take the optical image of the slide with a LAS-4000 mini luminescent image analyzer (ca. 30-s integration). Analyze the optical image with a specific image analysis software (Multi Gauge v3.1).

3.3 Imaging of Apoptosis Signaling in Living Cells

1. Grow COS-7 cells in a microslide (μ-slide VI$^{0.4}$) until the cell density reach ca. 85 % confluency. Transiently transfect the cells with *p*P2 dissolved in a lipofection reagent (TransIT-LT1), where the ratio is [DNA–TransIT-LT1–Opti-MEM = 1 μL: 3 μL:100 μL], and then incubate them in a CO_2 incubator (5 % CO_2, 37 °C) for 16 h.
2. Replace the culture medium in the microslide with a 50-mM KI-added HBSS buffer carrying (1) vehicle (0.1 % DMSO), (2) 0.01 μM staurosporine (STS; apoptosis inducer), or (3) 1 μM STS plus 50 μM Z-VAD-FMK (apoptosis inhibitor) (*see* **Notes 14** and **15**). Incubate the slide in a CO_2 chamber for 20 min (or a maximum of 2 h) (*see* **Note 16**).
3. Remove the buffer or culture medium from the channels of the microslide and simultaneously inject fresh HBSS buffer with dissolved nCTZ (nCTZ–HBSS = 1:100) through a multichannel pipet (*see* **Note 17**).

4. Determine the optical image with an image analyzer (LAS-4000) with ca. 5-s integration (*see* **Note 18**).
5. Analyze the optical image with a specific image analysis software (Multi Gauge v3.1).

3.4 Locating a Guest Luciferase in the PM of Living Cells

1. Grow COS-7 cells in a 35-mm glass-bottom plate until ca. 85 % confluency in a CO_2 incubator (5 % CO_2, 37 °C). Transiently transfect the cells with *p*P3 dissolved in a lipofection reagent (TransIT-LT1), where the ratio is [DNA–Opti-MEM–serum-free medium = 1 μL:3 μL:100 μL], and then incubate them in a CO_2 incubator (5 % CO_2, 37 °C) for 16 h.
2. Replace the culture medium in the plate with HBSS buffer carrying nCTZ, where the ratio is [nCTZ–HBSS = 1:100].
3. Determine the bioluminescence spectra with a high-sensitivity charge-coupled device (CCD) spectrophotometer (AB-1850) at 0 and 10 min after medium replacement (*see* **Note 19**). An integration time of 0.5–1 min is appropriate.
4. Normalize the spectra to the maximal intensity, e.g., in percentage (%).

3.5 Locating a Fluorescent Protein in the PM of Living Cells

1. Grow COS-7 cells in a 35-mm glass-bottom dish until ca. 85 % confluency. Transiently transfect the cells with *p*P4, where the ratio is [DNA–lipofection agent–serum-free medium = 1 μL:3 μL: 100 μL], and then incubate them in a CO_2 incubator (5 % CO_2, 37 °C) for 24 h (*see* **Note 20**).
2. Initiate apoptosis signaling of the cells by replacing the culture medium with an HBSS buffer-carrying vehicle (0.1 % DMSO) or 5 μM STS for 20 min.
3. Replace the HBSS buffer in the dish with fresh HBSS buffer.
4. View the cell dish under a fluorescent microscope (BX51) equipped with a band-pass filter at 510–560 nm excitation and with a long-pass filter at over 590 nm emission (*see* **Note 21**).
5. Take optical images in the automatic exposure mode and analyze them with a specific control software (AxioVision v4.6).

4 Notes

1. A secretion signal peptide (SP; 1–17 AA) is essential for bioluminescent capsules because copepod luciferases comprise multiple disulfide bonds that exert poor folding efficiency in the cytosol; thus, the ER is preferred for the folding. The SP is cleaved off and the host capsule is entering the secretory pathway. The SP-cleaved capsule is finally anchored at the PM

because of the MLS. This secretion nature is suppressed with an ER localization peptide, e.g., KDEL [10] or an MLS.

2. Any copepod luciferase may be replaced with Monsta in the bioluminescent capsule. As a bioluminescent leader, copepod luciferases are advantageous over other luciferases because they have the smallest molecular weight (20–24 kDa) and busting optical intensity.
3. The primer design is important for the efficient generation of cDNA blocks. Check the conventional rules on designing primers (in brief, refer to Fig. 2), which are a rough combination of educated guesses and old-fashioned trial and error.
4. Flexible linkers are important ingredients in constructing the bioluminescent capsule. The linkers generally consist of glycine (G) and serine (S) for flexibility and appropriate hydrophilicity. In the present study, a minimal length of the flexible linkers (no longer than 5) is appropriate between the domains. The shot linkers contribute to a dramatic conversion of the optical signal in the presence of an apoptosis signal.
5. Any guest proteins with specific functionality may be portable to the PM by the capsule.
6. The peptides “DEVD” and “IETD” should be inserted into the hinge region as substrates of apoptosis-signaling caspases. The peptides, “DEVD” and “IETD”, are the substrates of caspase 3 and 8, respectively. The peptides also act as a linker in the interface region.
7. Several membrane localization signals have been suggested by previous studies. The present signal peptide, “QGCMGLPCVVM”, is short but efficiently anchors the host capsule to the PM.
8. Take note of the frame shift in the codons. Double-checking the DNA sequence with the use of the sense and antisense primers is recommended for high fidelity.
9. Approximately 100–150 μL of the medium volume in each channel of the microslide is appropriate. The medium loss by evaporation is rapid on the microslide. The author recommends to replace the medium with a fresh one every 12 h. This unique microslide helps minimize the medium-driven autoluminescence owing to the minimal dead volume of the culture medium.
10. The recommended incubation time is around 16 h (overnight incubation). Researchers need not express the capsules to the full extent in the cells before experiment. TransIT-LT1 (Mirus) is a convenient lipofection reagent because no changing of medium is required after the transfection owing to its low cell toxicity.

11. The use of HBSS buffer is recommended for efficient live-cell imaging with suppressed autoluminescence.
12. Right before the imaging experiment, mix the HBSS buffer stored at room temperature with nCTZ (×100) kept at −20 °C (the ratio is [nCTZ–HBSS= 1:100]); Conduct all the experiments with reagents, which reach a constant room temperature for the stable luciferase reaction.
13. "Simultaneous" injection of the substrate solution with a multichannel pipet is important for precise comparison of the optical intensities.
14. A halide additive, KBr or KI, enhances the bioluminescence intensity.
15. The apoptosis inducer STS first activates both caspases 2 and 9 by inhibiting protein kinases, leading to caspase 3 activation and cleavage of the specific DEVD substrate in the bioluminescent capsule [11].
16. In case of long time stimulation of the cells with the apoptosis inducer (e.g., 2 h), the use of culture medium is appropriate instead of a HBSS buffer.
17. As an alternative method to fully suppress autoluminescence, the use of HBSS buffer with dissolved ViviRen (Promega) may be useful.
18. The integration time may be variable according to the cell density and buffer condition. The researchers may estimate an optimal exposure time using an auto-exposure mode of LAS-4000.
19. Many spectrophotometers have poor sensitivity to weak luminescence, especially in the near-infrared region.
20. In contrast to luciferases, fluorescent proteins require a longer fluorophore maturation time. Therefore, the incubation time should be extended according to the known maturation time of the guest fluorescent protein.
21. The excitation and emission filters should be adapted to the guest fluorescent protein.

Acknowledgments

This work was supported by a Grant-in-Aid for Young Scientists (B) from the Japan Society for the Promotion of Science (JSPS) [Grant number 23750096] and by a Grant-in-Aid for an A-Step program from the Japan Science and Technology Agency (JST) [Grant number AS232Z02075F].

References

1. Kim SB, Tao H, Umezawa Y (eds) (2009) Methods of analysis for imaging and detecting ions and molecules. In: R. Jelinek ed. Cellular and Biomolecular Recognition, Wiley-VCH: Darmstadt p 299
2. Loening AM, Dragulescu-Andrasi A, Gambhir SS (2010) A red-shifted Renilla luciferase for transient reporter-gene expression. Nat Methods 7(1):5–6. doi:10.1038/Nmeth0110-05
3. Li XY, Nakajima Y, Niwa K et al (2010) Enhanced red-emitting railroad worm luciferase for bioassays and bioimaging. Protein Sci 19(1):26–33. doi:10.1002/Pro.279
4. Kim SB, Ito Y, Torimura M (2012) Bioluminescent capsules for live-cell imaging. Bioconjug Chem 23:2221–2228. doi:10.1021/bc300323x
5. Alberts B (1994) Molecular biology of the cell, 3rd edn. Garland Pub, New York
6. Kim SB, Suzuki H, Sato M, Tao H (2011) Superluminescent variants of marine luciferases for bioassays. Anal Chem 83(22):8732–8740. doi:10.1021/Ac2021882
7. Kim SB, Ozawa T (2010) Creating bioluminescent indicators to visualise biological events in living cells and animals. Supramol Chem 22(7–8):440–449. doi:10.1080/10610278.2010.485251
8. Loening AM, Wu AM, Gambhir SS (2007) Red-shifted Renilla reniformis luciferase variants for imaging in living subjects. Nat Methods 4(8):641–643. doi:10.1038/nmeth1070
9. Ando Y, Niwa K, Yamada N et al (2007) Development of a quantitative bio/chemiluminescence spectrometer determining quantum yields: re-examination of the aqueous luminol chemiluminescence standard. Photochem Photobiol 83(5):1205–1210. doi:10.1111/j.1751-1097.2007.00140.x
10. Tannous BA, Kim DE, Fernandez JL et al (2005) Codon-optimized Gaussia luciferase cDNA for mammalian gene expression in culture and in vivo. Mol Ther 11(3):435–443. doi:10.1016/j.ymthe.2004.10.016
11. Caballero-Benitez A, Moran J (2003) Caspase activation pathways induced by staurosporine and low potassium: role of caspase-2. J Neurosci Res 71(3):383–396. doi:10.1002/Jnr.10493

Part III

Imaging of Disease (Pathogenesis)

Chapter 11

The Bioluminescent Imaging of Spontaneously Occurring Tumors in Immunocompetent *ODD-Luciferase* Bearing Transgenic Mice

Scott J. Goldman and Shengkan Jin

Abstract

The imaging of spontaneously occurring tumors in mice poses many technical and logistical problems. Recently a mouse model was generated in which a chimeric protein consisting of HIF-1α oxygen-dependent degradation domain (ODD) fused to luciferase was ubiquitously expressed in all tissues. Hypoxic stress leads to the accumulation of ODD-luciferase in the tissues of this mouse model which can be identified by noninvasive bioluminescence measurement. Crossing this transgenic mouse with tumorigenic mice yields solid tumors with hypoxic cores that may be successfully imaged and characterized using the technique described herein.

Key words ODD-luciferase, Bioluminescence, In-vivo, Hypoxia, Tumor, HIF

1 Introduction

Since the mid twentieth century, when the discovery of X-radiation and its development as a diagnostic aid provided the first opportunity to noninvasively image a living being, the detection and evaluation of tumors in vivo have been an area of ongoing research. In the latter half of the twentieth century, a wide array of noninvasive, live-tissue imaging options were developed to complement and even replace traditional X-radiography [1]. Rapid development and improvement in our understanding of ultrasonic imaging [2], computed tomography [3], positron emission tomography [4], magnetic resonance imaging [5], and nuclear scintigraphy [6] has provided us with a host of new tools to observe and measure tumors in vivo. These techniques provide access to a realm of detailed and important information, but to properly obtain and interpret this information, we must rely upon the expertise of board-certified radiologists within the medical profession who possess extensive training and experience.

Christian E. Badr (ed.), *Bioluminescent Imaging: Methods and Protocols*, Methods in Molecular Biology, vol. 1098, DOI 10.1007/978-1-62703-718-1_11,

In the laboratory, efforts are constantly underway to improve our understanding of tumor biology. To this end, scientists have developed a variety of tools, including the use of immunocompromised mice as hosts for human xenografts, thus providing researchers a method with which they may observe the growth and development of human tumors within the complex biological milieu of a living animal. These models, while tremendously useful, also lack some key properties that would allow them to better represent the development of tumors in their natural environment [7]. For instance, often xenografts behave in a different manner when transplanted in mice than they would in their native host [8, 9].

The recognized deficiencies in xenograft models have resulted in efforts to develop improved murine tumor models and imaging systems. Advances in genetic manipulation provide us with the ability to selectively knockout tumor suppressor genes and to overexpress oncogenes, thus allowing scientists to generate tumors in specific tissues which mimic the genetic changes found in human cancers [9–13]. Although murine tumors, whether induced or naturally occurring, are not always perfect analogs of their human counterparts, they provide valuable information on tumor behavior. In addition, they can be excellent systems for determining the efficacy of many anti-cancer therapeutics. One example of murine tumor models successfully used to this end is the development of transgenic mice predisposed to the development of mammary tumors [11, 14, 15].

While the development of transgenic mice has provided us with excellent models of stochastic, spontaneous tumorigenesis and development in vivo, the noninvasive imaging techniques available to observe these tumors remain largely inadequate. Human imaging modalities, including ultrasonography, computed tomography, and magnetic resonance imaging, often pose technical, financial, and regulatory challenges to the typical scientist. As a result, the scientific community has come to rely on more "user friendly" alternatives, the most popular of which is the use of bioluminescent or fluorescent imaging. Many effective in vivo imaging models have now been developed which depend either upon the conditional co-expression of bioluminescent reporter genes with specific tumor-associated genes, or upon the use of fluorescent reporter molecules directed towards specific cancer-associated cell proteins [16]. These models, however, are inherently limited in use to a single type or subtype of cancer, and may not be broadly applied to tumors of differing origins.

With the model described here, we attempted to alleviate many of the deficiencies of murine mammary tumor models by combining the ease of use of bioluminescent imaging with the superior biological model of stochastically occurring in vivo tumors in immunocompetent mice. This method, as with all tumor imaging techniques, depended upon the differentiation of tumor cells from normal tissue. It has long been known that there are differences between cancerous tissue and normal tissue that, if exploited, might provide a foundation for a more broadly applicable tumor imaging system [17]. For instance, all solid tumors have been shown to experience a certain degree of relative hypoxia when compared with surrounding tissue [18–20].

Recently, a transgenic mouse expressing the oxygen-dependent degradation (ODD) domain of the Hypoxia Inducible Factor 1-α (HIF1-α) gene fused to a luciferase bioluminescent reporter gene was developed by Safran et al. [21]. This model allows for the bioluminescent imaging of regions in which hypoxia is present. The HIF1-α gene is ubiquitously transcribed and translated, but under normoxic conditions, the enzyme HIF prolyl hydroxylase utlilizes oxygen to hydroxylate the HIF ODD. The hydroxylated ODD then recruits von Hippel–Lindau protein (pVHL), a ubiquitin E3 ligase, leading to poly-ubiquitination of HIF1-α and its subsequent degradation by the proteasome. Conversely, under hypoxic conditions, lack of hydroxylation of the ODD domain leads to the accumulation of HIF1-α which acts as a transcriptional regulator for a number of hypoxia-induced genes. By universally expressing the ODD-luciferase gene in mice, Safran et al. developed an organism in which luciferase would be continuously transcribed and translated in all cells, but would also be rapidly degraded under normoxic conditions. Only under conditions of hypoxia can luciferase accumulate to the point that it can act upon the substrate luciferin and generate a bioluminescent signal that may be imaged.

Our technique is based upon the hypothesis that solid tumors should be visible in *ODD-luciferase* expressing mice due to the relative hypoxia experienced by the tumor cells, essentially allowing for the use of the *ODD-luciferase* transgenic model as a platform for noninvasive imaging of spontaneous solid tumors in mice [22]. For our work, we utilize mice predisposed to the development of mammary gland tumors (*MMTV-neu* and *Beclin1* +/−). These mice were cross-bred with *ODD-luciferase* mice. The resultant cross-breeds yielded tumors which generated substantial bioluminescent signals clearly discernable from background tissue luminescence. This technique may be used to track tumor growth and development longitudinally over several weeks.

2 Materials

2.1 Anesthesia

1. Compressed oxygen—provided either by tank or via central building supply.
2. Isoflurane anesthetic (Forane) (Baxter Healthcare, Deerfield, Ill.)
3. IVIS Spectrum II workstation (Caliper Life Sciences, Hopkinton, MA), with integrated oxygen flow meters, isoflurane vaporizer, and induction chamber.

4. Tumor-bearing ODD-luciferase transgenic mice (ODD-luciferase mice are available from Jackson Labs, Bar Harbor, ME as FVB.129S6-*Gt(ROSA)26Sor*$^{tm1(HIF1A/luc)Kael}$/J, stock number 006206) (*see* **Note 1**).

2.2 Luciferin Preparation and Injection

1. VivoGlo Luciferin, In Vivo Grade (D-luciferin may also be used. All substrates hereafter will be referred to simply as "luciferin").
2. Sterile Phosphate Buffered Saline (PBS).
3. 500 ml sterilized glass beaker.
4. Ice with container.
5. Sterile 10 ml tubes.
6. –20 °C Freezer.
7. Sterile 60 cc Luer Lock syringe.
8. 22 μm syringe filter.
9. Aseptic tissue culture hood, properly installed.
10. Stock solution of 30 mg/ml luciferin in PBS, frozen.
11. Sterile Phosphate Buffered Saline.
12. Ice with container.
13. Sterile 10 ml tubes.
14. –20 °C Freezer.
15. Sterile 10cc Luer Lock syringe.
16. Aluminum foil.
17. Aseptic tissue culture hood, properly installed.
18. Gram scale (for weighing mice).
19. Prepared solution of 3 mg/ml luciferin in sterile PBS, thawed on ice in light-shielded tube.
20. Sterile 1 ml syringe.
21. Calculator (if needed).
22. 26 g needle.

2.3 Initial Imaging

1. IVIS Spectrum II in vivo imaging station (or equivalent) with LivingImage Software, version 3.2 or later (Caliper Life Sciences, Hopkinton, MA).
2. Anesthetized tumor-bearing ODD-luciferase transgenic mice (ODD-luciferase mice are available from Jackson Labs, Bar Harbor, ME as FVB.129S6-*Gt(ROSA)26Sor*$^{tm1(HIF1A/luc)Kael}$/J, stock number 006206).

2.4 Post-imaging Recovery

1. Standard mouse cage with wire top.
2. Heat source (heating pad, heat lamp, or other thermostatically controlled heat source).

2.5 Image Processing

1. Windows-based personal computer.
2. LivingImage software, version 3.2 or later (Caliper Life Sciences, Hopkinton, MA).

2.6 Depilation

1. Depilatory cream of choice.
2. Small animal electric hair clippers (battery or AC powered).
3. Cotton towels, pledgets, towelettes, or "4 × 4" surgical sponges.

3 Methods

3.1 Anesthesia

For all imaging procedures, anesthesia was limited to inhalational isoflurane. Isoflurane is the preferred anesthetic for both induction and maintenance due to the reduced stress on the animal (*see* **Note 1**).

1. Prepare IVIS imaging station for use. This included switching on the apparatus, waiting for the unit to self-calibrate, and ensuring that the heating pad within the imaging station is warm via both temperature read-out and tactile confirmation.
2. Select mice for induction (*see* **Note 2**).
3. Remove selected mice from cages and place the mice in an induction chamber connected to an isoflurane vaporizer.
4. Turn on the oxygen and set the isolation chamber flow-meter at 1 L/min.
5. Set the isoflurane vaporizer at 3 % (*see* **Note 3**).
6. Observe mice for signs of general anesthesia, notably cessation of movement and decreased respiratory rate (*see* **Note 4**).
7. When all mice have been successfully anesthetized (2 min after the last mouse has ceased moving), set the isoflurane vaporizer to 0 % and flush the induction chamber with 1 L/min of pure oxygen for 5 s (*see* **Notes 5** and **6**)
8. Remove mice for injection of luciferase solution (see Subheading 3.2 below) *see* **Notes 7–9**.

3.2 Luciferin Preparation and Injection

1. Turn out or black out all nonessential light sources in the room. All procedures should be carried out in a darkened aseptic tissue culture hood (*see* **Note 10**).
2. Place 330 ml of cold PBS, pH 12, in a 500 ml glass beaker.
3. Take 10 mg of luciferin solid and dissolve in the 330 ml of sterile PBS, gently swirling to mix, to yield a solution of 30 mg/ml luciferin.
4. Filter the 30 mg/ml solution by drawing the solution into a 60cc syringe and then forcing the solution through a 0.22 μm

filter into sterile, pre-labelled 10 ml canonical tubes. Place tubes in a −20 °C freezer for storage.

5. Turn out or black out all nonessential light sources in the work room. All procedures should be carried out in a darkened aseptic tissue culture hood (*see* **Note 10**).
6. Take one 10 ml tube of 30 mg/ml luciferin from the −20 °C freezer and immediately wrap with aluminum foil to shield the solution from ultraviolet light.
7. Thaw the 30 mg/ml luciferin by warming in your hands, then place on ice.
8. Fill a 10 ml tube with 9 ml PBS.
9. Draw 1 ml of 30 mg/ml luciferin solution by pipette or syringe, and add to the prepared tube of PBS to create a solution of 3 mg/ml luciferin in PBS.
10. Cover the tube with aluminum foil, mix gently, and place on ice for immediate use.
11. Return the 30 mg/ml luciferin stock to the −20 °C freezer for storage.
12. Determine weights of mice (*see* **Note 11**).
13. Calculate the amount of luciferin needed for injection. In this protocol, a dose of 50 mg/kg was used (*see* **Note 12**).
14. Draw up needed amount of luciferin in a 1 ml syringe with a small (26 g) needle (*see* **Notes 13** and **14**).
15. Remove the candidate mouse from the induction chamber and scruff by the neck between thumb and forefinger of the non-dominant hand while using the pinky finger of the same hand to secure the base of the tail Hold the mouse head-down at a 45° angle—the belly should be exposed (*see* **Note 15**).
16. Identify the midpoint between the xiphoid and the pelvic bone, then insert the needle at a 45° angle to the mouse using your dominant hand, bevel-up, slightly off-center from the mid-point. Insert needle to a depth of approximately 8–12 mm (*see* **Note 16**).
17. Depress syringe plunger and inject the luciferin solution.
18. Withdraw the needle from the abdomen, observe briefly for bleeding or oozing of fluid from injection site (*see* **Note 17**).
19. Immediately following the luciferin injection, place the mouse in the imaging chamber on the heating pad. Remove the stopper from the anesthesia nose cone in the imaging chamber and position the mouse as desired for imaging with the nose in the anesthesia cone/block (*see* **Note 18**). Turn on the oxygen flow meter for the imaging chamber and set to a flow rate of 1 L/min and set the isoflurane vaporizer to 1.5 % Repeat for all mice used in an imaging session.

3.3 Initial Imaging (See Note 19)

1. Click on the living image software icon on the computer desktop.
2. When the software opens, click "Initialize."
3. Wait for the "Temperature" box in the LivingImage software control panel to turn green.
4. indicating the CCD (Charge-coupled device) image sensor has reached the appropriate temperature (*see* **Note 20**).
5. Ensure the check boxes next to the "Luminescent" and "Photograph" headings are checked.
6. Confirm that the excitation filter setting is set to "Block" and the emission filter is set to "Open" (*see* **Note 21**).
7. Set Exposure Time to "Auto" for initial imaging.
8. Select your field-of-view (FOV) from the drop-down list (*see* **Note 22**).
9. Set the focus by selecting "use subject height" from the focus drop-down list (*see* **Note 23**).
10. Set the photographic image setting to "Auto."
11. When ready to acquire the image, click "Acquire."
12. Upon acquisition of the image, observe the color bar and determine if your signal of interest is above the background noise level and below the threshold of CCD saturation.
13. Adjust exposure or binning as necessary to bring the signal of interest within range.
14. Repeat from **step 1** as necessary (*see* **Notes 24** and **25**).

3.4 Post-imaging Recovery

1. Turn off the oxygen flow meter to the imaging chamber.
2. Remove mice from imaging chamber and immediately place in a heated recovery cage (*see* **Note 26**).
3. Observe mice for full recovery from anesthesia, indicated by normal movement and mentation.
4. Return the mice to their regular housing cages after no less than 30 min of observation.

3.5 Image Processing

1. Click the "Browse" button on the LivingImage software and select the data of interest. The selected data will be displayed in the browser.
2. Select the desired images for the composite image (*see* **Note 27**).
3. Click "Load as Group" to open multiple images within a single window. This function will force all images to conform to the same intensity scale. This allows the researcher to compare image intensity among multiple views of a single mouse.
4. Open an image or image sequence and click "ROI tools" in the tool palette. Select an image in which the tumor is readily

visible from either a lateral, dorsal, or ventral perspective (*see* **Note 28**).

5. Click the ROI "contour" shape button.
6. Use the ring that appears to identify the location of your ROI.
7. Click on the "Create" button on the ring to obtain area and signal intensity.
8. Repeat **steps 1** through **4** for a secondary view of the same tumor (i.e., two views used for a tumor of the left mammary pad might be right lateral recumbency and dorsal recumbency) (*see* **Note 29**).
9. Utilize the two cross-sectional areas to determine tumor volume via the equation for the volume of an ellipsoid: $V = 4/3 \times \pi \times a \times b \times c$, where a is the radius of the x axis, b is the radius of the y axis, and c is the radius of the z axis (*see* **Note 30**).

3.6 Optional Subject Depilation for Improved Imaging (*See* Notes 31 and 37)

3.6.1 Depilatory Cream Method

1. Anesthetize the subject utilizing the method described above in Subheading 3.1.
2. Flush the induction chamber with oxygen for 30 s, then open the chamber and remove the subject.
3. Working rapidly, apply a thin layer of depilatory cream to the subject (*see* **Note 32**).
4. Replace the subject in the induction chamber and reinitiate anesthesia for the duration of time indicated for the depilatory being used.
5. Flush the induction chamber with oxygen and remove the subject.
6. Working quickly, remove the cream with a clean cloth and then wipe the area down with a damp, clean cloth to remove residual cream residue (*see* **Note 33**).
7. Place the subject on the heated imaging stage in the IVIS machine and initiate maintenance anesthesia per instructions in Subheading 3.1 (*see* **Notes 34** and **35**).

3.6.2 Mechanical Clipping Method

1. Anesthetize the subject utilizing the method described above in Subheading 3.1.
2. Flush the induction chamber with oxygen for 30 s, then open the chamber and remove the subject.
3. Working rapidly but carefully, use well-oiled and sharpened small-animal hair clippers to remove the hair from the area of interest.
4. Place the subject on the heated imaging stage in the IVIS machine and initiate maintenance anesthesia per instructions in Subheading 3.1 (*see* **Note 36**).

4 Notes

1. Be aware that some mouse breeds may be resistant to some tumor transgenes (for example: FVB mice are resistant to the MMTV-neu transgene, which increases the chances of mammary carcinomas in mice).
2. Among the inhalant anesthetics, isoflurane places the least amount of stress on the liver. Only 0.2 % of isoflurane is metabolized via the liver, versus 20 % for halothane and 3 % for sevoflurane. Additionally, inhalational anesthesia via the use of induction chambers significantly reduces the risk of trauma associated with restraint and injection of injectable anesthetics, provides a more measured and gentler induction of anesthesia, is more cost effective than the use of injectables, typically results in fewer adverse events, and is rapidly reversible.
3. Great care should be taken when selecting candidate mice that the means for positively identifying one mouse from another is maintained. Often, mice are identified by cage number and ear punch, or some similar technique. When combining mice from different cages (a common occurrence when imaging multiple cohorts), it is quite easy to accidentally mix mice with identical ear-punches from different cages. These mistakes must be avoided at all costs if useful data is to be obtained.
4. When inducing anesthesia in mice via the use of isoflurane induction chambers, concentrations up to 5 % isoflurane provided in a flow rate of 1 L/min oxygen are commonly used. Different breeds, genotypes, and ages of mice will react differently to anesthetic induction, and the investigator must take care to properly monitor the mice during induction and adjust the isoflurane concentration to an appropriate level. In our experiments, 3 % isoflurane provided in 1 L/min oxygen provided excellent results.
5. Some mice, most notably C57Bl6 breeds, exhibit agonal breathing upon anesthetic induction with isoflurane. This condition often causes researchers to increase the gas percentage in an effort to combat the apparent waking from anesthesia. However, increasing the depth of anesthesia often causes agonal breathing—and the cycle continues until the mice die from anesthetic overdose. The moral of this story is that respiratory rate and pattern should not be used as the sole indicator of successful anesthesia in mice.
6. The induction chamber should be flushed for the safety of the researchers prior to opening the chamber. Excessive inhalation of isoflurane can induce significant headaches.
7. Care must be taken to move rapidly once mice are removed from the induction chamber, as they typically awaken 3–5 min

after the discontinuation of isoflurane exposure. If the mice do recover, they may be re-anesthetized in the induction chamber before continuing the protocol.

8. All researchers should be aware of the difference between anesthesia (induced loss of consciousness with amnesia) and analgesia (mitigation of pain). Isoflurane anesthesia provides the former, but not the latter. The use of invasive or otherwise painful procedures necessitates the use of analgesics, typically before, during, and after anesthesia.

9. During anesthesia, body temperature can drop rapidly in mice. It is therefore essential that anesthetized mice be placed in a heated environment. During imaging, this is accomplished via the built-in heating element in the imaging platform. Before and after imaging, this may be accomplished with heated cages, heating pads, or heat lamps. Whichever heat source is used, it is critical to monitor mice under or recovering from anesthesia to prevent hyper or hypothermia. It is also worth noting that keeping the mice warm (but not overheated) will increase peripheral vasodilation and thus decrease the background signal often present in anesthetized ODD-Luciferase mice. The peripheral vasoconstriction typically present under anesthesia and exacerbated by environmentally induced hypothermia can generate significant noise by increasing peripheral tissue hypoxia in these mice.

10. Volatile anesthetics (and injectables) have been shown to have an inhibitory effect on luciferase [23].

11. Reconstituted luciferin must be stored at –20 °C and protected from light exposure. UV exposure will result in the conversion of luciferin to dihydroluciferin, a competitive inhibitor of the luciferase/luciferin reaction.

12. Mouse weights are best obtained prior to anesthesia and imaging procedures, but may be obtained at this time if necessary. As a matter of course, study mice should be weighed on a weekly basis by either laboratory or vivarium staff, negating the need for this step.

13. In this protocol, we utilized a 50 mg/kg dose of luciferin. Typical doses can range from 50 mg/kg up to 150 mg/kg. Varying metabolic rates among individual mice can result in considerable differences in signal strength.

14. If the weights of subject mice area available ahead of time, significant time savings may be realized by preparing multiple customized syringes for dosing the individual mice. One option is to tailor the amount of luciferin solution in each syringe to an individual mouse, while the other option is to draw up identical volumes into multiple syringes and then customize the dose by injecting only the appropriate amount. In either case,

pre-drawn syringes should be maintained on ice and covered with a light-proof barrier.

15. When drawing up doses into syringes, it is appropriate to draw the solution into the syringe via the needle that will be used during the injection. This technique allows the dead space in the needle hub to be filled with luciferin solution, thus ensuring the appropriate dose is provided to the mouse during injection when the volume is measured off the syringe markings.
16. Holding the mouse this way shifts the abdominal organs under the rib cage and assists in the prevention of "needle stoppers" (accidentally inserting the needle into solid organs) and intra-luminal injections (injections into the interior space of the stomach or intestines).
17. If at any time you feel resistance during the injection process, immediately stop advancing the needle and partially withdraw the syringe. It is appropriate to slightly redirect the needle if necessary and then perform the injection. If there are concerns about advancing the needle too far, a needle-guard can be fashioned by clipping 10 to 12 mm off the tip of a needle cap and then placing the tipless cap over the needles used for injection. This apparatus will block the advance of the needle any further than 1 cm under the skin, thus alleviating the risk of accidents. Another solution is to utilize tuberculin syringes for the injections, which possess very short, integrated needles that cannot be advanced more than 5 mm under the skin.
18. Occasionally, you may see a droplet of luciferin solution exude from the injection site. This is usually an indication that the injection was not peritoneal, but perhaps intra-adipose, intramuscular, intradermal, or even subcutaneous. In these situations, it is unlikely that a useable signal will be obtained during imaging, but there is little harm in attempting to image the mouse regardless. Do not attempt to reinject when this occurs. If anything other than luciferin solution is observed exuding from the injection site (blood, peritoneal fluid, urine, etc.), the vivarium staff must be notified immediately so that proper care may be provided, and the incident will likely need to be recorded as an "adverse incident" per local IACUC guidance. See your local vivarium director or lab animal veterinarian for further guidance.
19. Take care when placing the nose into anesthesia delivery cone/block that the fit is not excessively tight. This can result in dyspnea or even death.
20. These instructions are based on the LivingImage software version 3.2.
21. The CCD must be cooled to between −90 °C and −105 °C for a safe and effective operation.

22. The excitation filter is used to block or allow the excitation fluorescence generated by the IVIS machine for fluorescence imaging. The emission filter allows for the recording of bioluminescent and fluorescent images.
23. The IVIS machine and software allow for a limited number of pre-set fields-of-view. These roughly accommodate anything from a small subsection of a mouse to five complete mice. Note that a smaller FOV will result in higher sensitivity when imaging. Be wary of using small FOV's, however, as the FOV is controlled by moving the stage up and down within the imaging chamber, and selecting an FOV too small may result in injury to the animals or the machine.
24. The average mouse will have a height above the stage of approximately 5 cm. The use of Manual focus is not recommended.
25. Luciferin metabolism differs among individuals and also differs within the same individual depending on a variety of factors, including heart rate, temperature, and respiratory rate. To identify peak luciferin uptake resulting in the best images it is recommended that, at least for the first several trials, multiple images are acquired for each mouse over time. Utilizing this method, it is possible to determine roughly when peak luciferin availability occurs and to adjust the timeline of future experiments to take advantage of this. The "Acquire Sequence" function of the LivingImage software allows the researcher to arrange for a series of timed images to be acquired over a given period (for example: one shot every 60 s for 15 min) and provides the most effective way to determine peak signal intensity among subjects.
26. Background signals in mice universally expressing the *ODD-Luciferase* transgene will routinely cause problems in mice with small (<3 mm diameter) tumors. Areas of normal background signal generation include the thyroid, the kidneys, the extremities (especially when cold), and the ventral mammary fat pads. While there appears no way to mitigate the background signals generated by the chronically and physiologically hypoxic thyroid and kidneys, the mammary fat pad signals seem to be the result of peripheral vasoconstriction common during anesthesia combined with a transient ischemia caused by protracted ventral recumbency during anesthesia and imaging. The mammary fat pad background signal can be reduced by maintaining the mouse under anesthesia in a warm environment in dorsal recumbency for several minutes. Additionally, when tumors reach a certain size (>5 mm), the signal observed becomes increasingly intense, and that intensity will drive background signals, including the comparatively strong kidney and thyroid signals, progressively farther towards the lower end of the

scale, allowing for better differentiation of the tumor from surrounding tissues.

27. Mice are prone to rapid hypothermia during and following anesthesia due to their very small surface to volume ratio. To counteract hypothermia, it is necessary to maintain the mice in a heated environment during recovery. Care must be taken to prevent the environment from being too hot, however, as anesthetized animals cannot thermoregulate effectively and may be subject to damage from hyperthermia or burns if care is not taken.
28. Selecting multiple images from the LivingImage browser window can be a daunting task, as thumbnails of the images are not available and the only way to identify an image is to open it completely. The difficulty of this task can be mitigated by providing as much information as possible about an image or image sequence at the time of imaging. The more data available, the easier it is to identify individual images of interest via the browser. The alternative is to open multiple images to identify those of interests, note their file names, and then reopen the browser and select those images to be opened as a group in a composite. As with all tasks, meticulous record-keeping will facilitate future use of collected data.
29. To determine tumor cross-sectional area, only one image of the tumor is required. The best image is one in which the tumor is closest to the CCD, but certain conventions must be employed for uniformity of data. To this end, mice in this study were imaged in dorsal recumbency, ventral recumbency, and right and left lateral recumbency. The optimal tumor image for cross-sectional area determination was obtained from the image in which the tumor was closest to the CCD (i.e., a tumor on the ventrum would be measured using the dorsal recumbent view).
30. The automatic ROI drawing method is the preferred method for use, as it minimizes variability in the determination of ROI borders among different subjects. However, some subject may exhibit a background level of bioluminescence prohibitive to the use of the automatic ROI tool and will require instead the use of a "free-draw" ROI.
31. In any equation for the volume of an ellipse, there will be three radii to account for. The radius of the long axis of the tumor should be identical for both views (lateral and dorsal or ventral) for this equation to provide meaningful data. This process is simplified by the LivingImage software cross-sectional area function, which provides x and y axis length for each ROI.
32. The methods described herein were all used on mice with white hair coats. Mice with agouti, brown, and black hair coats were

also imaged, but colored coats profoundly attenuated the signal strength. This effect was not always negative, as in cases of tumor growth where the hair coat attenuated all background signal, allowing only the stronger tumor signal to be observed. However, for the collection of accurate luminescence strength data, the hair on these animals should be removed prior to imaging.

33. Depilatory creams are irritants, even in small amounts. Take great care to apply only to the area of interest, and to strictly observe the guidelines for length of contact. Misapplication or misuse may result in blindness, loss of vibrissae, and severe dermal hyperemia which can confound imaging results.

34. For this task, paper towels are completely inadequate. The researcher should have on-hand either cotton "4 × 4" surgical sponges or cotton towels/towelettes available.

35. When deprived of their hair coat, dampened, and subjected to anesthesia, mice are at profound risk of hypothermia and special care must be taken to keep their body temperature in a comfortable range.

36. Following the manipulations associated with hair removal, it is common for mice begin to arouse from anesthesia. If mice are "light" and beginning to waken, they may need additional time under induction levels of anesthesia either in the induction chamber or in the imaging chamber (but only in the latter if they remain immobile/non-ambulatory).

37. While either depilation method may be effective alone, it may be necessary to combine both methods, with mechanical shaving followed by chemical depilation, for best results. If this method is utilized, the mechanical shaving must be carried out scrupulously to avoid skin abrasion and subsequent severe irritation by the cream.

References

1. Burns S (2004) Oncology: tumors and treatment—a photographic history, 1845–1945. Burns Archive Press, New York, NY
2. Dalla Palma L, Pozzi Mucelli RS, Ricci C, Zuiani C (1989) Ultrasonography in oncology. A review. Acta Oncol 28:157–162
3. Bragg DG (1977) Advances in diagnostic radiology: problems and prospects. Cancer 40: 500–508
4. Hawkins RA, Phelps ME (1988) PET in clinical oncology. Cancer Metastasis Rev 7:119–142
5. Frank JA, Dwyer AJ, Doppman JL (1987) Nuclear magnetic resonance imaging in oncology. Important. Adv Oncol: 133–174
6. Mansfield CM, Park CH (1985) Contribution of radionuclide imaging to radiation oncology. Semin Nucl Med 15:28–45
7. Dennis C (2006) Cancer: off by a whisker. Nature 442:739–741
8. Sharpless NE, Depinho RA (2006) The mighty mouse: genetically engineered mouse models in cancer drug development. Nat Rev Drug Discov 5:741–754
9. Ince TA, Richardson AL, Bell GW et al (2007) Transformation of different human breast epithelial cell types leads to distinct tumor phenotypes. Cancer Cell 12:160–170
10. Shen Q, Brown PH (2005) Transgenic mouse models for the prevention of breast cancer. Mutat Res 576:93–110
11. Muller WJ (1991) Expression of activated oncogenes in the murine mammary gland: transgenic models for human breast cancer. Cancer Metastasis Rev 10:217–227

12. Leenders MW, Nijkamp MW, Borel Rinkes IH (2008) Mouse models in liver cancer research: a review of current literature. World J Gastroenterol 14:6915–6923
13. Jeet V, Russell PJ, Khatri A (2010) Modeling prostate cancer: a perspective on transgenic mouse models. Cancer Metastasis Rev 29:123–142
14. Guy CT, Webster MA, Schaller M et al (1992) Expression of the neu protooncogene in the mammary epithelium of transgenic mice induces metastatic disease. Proc Natl Acad Sci U S A 89:10578–10582
15. Muller WJ, Arteaga CL, Muthuswamy SK et al (1996) Synergistic interaction of the Neu proto-oncogene product and transforming growth factor alpha in the mammary epithelium of transgenic mice. Mol Cell Biol 16:5726–5736
16. Hwangdo W, Ko HY, Lee JH et al (2010) A nucleolin-targeted multimodal nanoparticle imaging probe for tracking cancer cells using an aptamer. J Nucl Med 51:98–105
17. Tennant DA, Duran RV, Gottlieb E (2010) Targeting metabolic transformation for cancer therapy. Nat Rev Cancer 10:267–277
18. Koh MY, Spivak-Kroizman TR, Powis G (2010) HIF-1alpha and cancer therapy. Recent Results Cancer Res 180:15–34
19. Michieli P (2009) Hypoxia, angiogenesis and cancer therapy: to breathe or not to breathe? Cell Cycle 8:3291–3296
20. Harris AL (2002) Hypoxia–a key regulatory factor in tumour growth. Nat Rev Cancer 2:38–47
21. Safran M, Kim WY, O'Connell F et al (2006) Mouse model for noninvasive imaging of HIF prolyl hydroxylase activity: assessment of an oral agent that stimulates erythropoietin production. Proc Natl Acad Sci U S A 103: 105–110
22. Goldman SJ, Chen E, Taylor R, Petrosky W et al (2011) Use of the ODD-luciferase transgene for the non-invasive imaging of spontaneous tumors in mice. PLoS One 6(3):e18269
23. Keyaerts M, Remory I, Caveliers V et al (2012) Inhibition of firefly luciferase by general anesthetics: effect on in vitro and in vivo bioluminescence imaging. PLoS One 7(1):e30061

Chapter 12

Blood-Based Assay with Secreted Gaussia Luciferase to Monitor Tumor Metastasis

Hiroshi Yamashita, Dan T. Nguyen, and Euiheon Chung

Abstract

Only a few techniques are currently available for quantifying systemic metastases in preclinical models. Cancer cell expression of naturally secreted Gaussia luciferase (Gluc) provides a useful circulating biomarker that enables the monitoring of metastatic tumor burden and of treatment response from minimal drops of blood. This blood-based Gluc assay exhibits several distinct advantages: (1) It is highly sensitive in quantifying metastatic tumor growth, particularly when compared to whole-body bioluminescence imaging (BLI) alone; (2) It is quantitative by nature and reflects viable tumor burden in a minimally invasive manner; (3) Through longitudinal collection of blood samples, treatment response can be monitored in real-time; and (4) Gluc bioluminescence provides a means to localize and assess metastatic colonization using BLI. By elucidating the progression of systemic metastases and therapeutic response in animal models, the blood-based Gluc assay is emerging as a valuable quantitative tool for novel drug development.

Key words Biomarker, Gaussia luciferase, Bioluminescence, Treatment response monitoring, Tumor metastasis, Blood-based assay

1 Introduction

The evaluation of metastatic tumor burden in animal models is not trivial. Oftentimes, metastasis can only be assessed at the sacrificial end point while the burden during disease progression remains unknown. This challenge is particularly problematic for preclinical therapeutic studies, where it is important to control for the tumor burden at the start of any treatment. Currently, bioluminescence imaging (BLI), paired traditionally with firefly luciferase (Fluc) as the reporter, is used to localize tumors and to evaluate metastatic tumor growth in vivo. Although BLI is a powerful tool in itself, the optical signal from Fluc, as well as those of other optical reporters, provides only a semi-quantitative measurement of metastatic tumor burden. A highly sensitive quantitative assessment is hindered not only by the relatively poor spatial resolution of BLI due to light scattering but also by the attenuation of an optical signal traveling

Christian E. Badr (ed.), *Bioluminescent Imaging: Methods and Protocols*, Methods in Molecular Biology, vol. 1098,
DOI 10.1007/978-1-62703-718-1_12, © Springer Science+Business Media New York 2014

through living tissue. These limitations make tumors located deep inside tissue appear very different than their true size, thereby inaccurately reflecting the burden of disease. For small metastatic cell clusters, evaluation becomes even more difficult, if not impossible [1]. For these reasons, it is impractical to employ BLI as a strict quantitative comparison of tumor burden between different experimental animals.

Secreted reporters in the blood have emerged as promising tools for the detection, quantification, and noninvasive monitoring of biological processes in experimental models [2–11]. Tumor cell expression of the naturally secreted Gaussia luciferase (Gluc), derived from the marine copepod *Gaussia princeps*, has been demonstrated as a sensitive and quantitative method for evaluating cancer progression in vivo [2, 3, 11–13]. Gluc exhibits advantages over other commonly used reporters for BLI. It is 20,000-fold more sensitive than secreted alkaline phosphatase and approximately 2,000-fold more sensitive than Fluc or *Renilla* luciferase (Rluc), although this measure may vary due to differences in peak emission wavelength (480 nm for Gluc; 612 nm for Fluc) and the depth of a particular tumor from a tissue surface [5, 14]. Furthermore, since Gluc is a secreted enzyme, its concentration in bodily fluids (i.e., whole blood, urine, plasma, etc.) correlates with its expression by a given biological system, thereby quantitatively reflecting the amount of viable cancer cells in primary and metastatic tumors [2, 3, 12]. The blood-based Gluc assay is highly sensitive and provides a novel means to longitudinally monitor metastatic progression as well as treatment response [2, 11].

Application of the Gluc reporter confers additional advantages. When tumor growth is at a stage that is not yet identifiable by BLI, the blood-based Gluc assay facilitates the detection of such early systemic metastasis. Also, Gluc acts on a different substrate (i.e., coelenterazine) than Fluc (i.e., d-luciferin) for bioluminescence; if the reporters are co-expressed, this enables simultaneous monitoring of both tumor burden and signaling pathway activity [15, 16]. As a whole, the quantitative nature of the Gluc reporter and the experimental capabilities have made Gluc assay technology an emerging tool in cancer research. It improves the study of tumor metastasis and aids in the development of strategies for treating this devastating disease.

2 Materials

2.1 Gluc Transfection

1. The MDA-MB-231 cell line (231BR): provided by Dr. Patricia S. Steeg (National Cancer Institute, Bethesda, MD) (*see* **Note 1**) [17].
2. Dulbecco's modified Eagle Medium (DMEM) with 10 % Fetal bovine serum (FBS) and 1 % penicillin–streptomycin (P/S) solution.

3. Gluc-CFP lentivirus: provided by Dr. Bakhos A. Tannous (Harvard Medical School, Boston, MA) (*see* **Note 2**) [3].
4. Polybrene ((hexadimethrine bromide).

2.2 Experimental Metastasis Model (Intracardiac Injection Model)

1. Cell strainer (40 μm pore size).
2. Phosphate Buffered Saline (PBS).
3. 6–7-week-old female nude mice.
4. Anesthetics.
5. 30 g needle.

2.3 Blood/Urine Gluc Assay

1. 50 mM EDTA (Ethylenediaminetetraacetic acid).
2. Scalpel.
3. Coelenterazine (CTZ, Nanolight, Pinetop, AZ).
4. Methanol mixed with few drops of HCl (hydrogen chloride) for reconstitution of CTZ.
5. 5 mM NaCl, pH 7.2.
6. Opaque (white or black) 96-well plate.
7. Plate luminometer (a Centro XS, Berthold Technologies, Oak Ridge, TN was used for bioluminescence imaging).

2.4 Bioluminescence Imaging

1. IVIS Imaging System (Lumina II, Caliper Life Sciences, Hopkinton, MA) was used for BLI recording.
2. Living Image software 3.0 coupled to the IVIS system.
3. 50 U insulin syringe (26 g needle).

3 Methods

3.1 Gluc Transfection

1. Culture 231BR cells in a 6-well plate in DMEM with 10 % FBS and 1 % P/S (10 % DMEM) at 37 °C.
2. When cells have grown to 60–70 % confluency, replace the medium with 1 ml of 10 % DMEM containing lentivirus at a multiplicity of infection of 50 and with 1 μl of 8 mg/ml polybrene. Continue cell culture for 24 h (*see* **Note 3**).
3. Replace the lentiviral medium with 10 % DMEM for continued cell culture.
4. After 2–3 days, check CFP (cerulean fluorescent protein) expression using fluorescence microscopy. If fluorescence is faint, repeat the transfection with the same dose of lentivirus (*see* **Note 4**).
5. Sort CFP positive cells by FACS (Fluorescence-activated cell sorting).

3.2 Experimental Metastasis Model (Intracardiac Injection Model)

1. Once cultured to 80 % confluency, trypsinize and resuspend the 231BR cells. Centrifuge the cell suspension at 500×*g* at 4 °C for 3 min and remove the supernatant. Rinse the cells with Phosphate Buffered Saline (PBS), centrifuge at 500×*g* at 4 °C for 3 min, and remove the supernatant. Repeat this step twice to remove the media.
2. Resuspend the cell pellet with PBS, and filtrate with a cell strainer to achieve single cell suspension. Dilute with PBS to a concentration of 2.5×10^6 cells/ml. Keep the cell suspension on ice for intracardiac injections (*see* **Note 5**).
3. Anesthetize the nude mice to inject 100 μl of cell suspension (2.5×10^5 cells) into the left ventricle with a 30 g needle (*see* **Notes 6** and 7) [17].
4. The needle is inserted vertically into the left ventricle. Proper insertion can be confirmed by the reflux of pulsing bright red blood into the syringe followed by slow injection. At the end of injection, pulsing blood reflux should be reconfirmed before needle retraction.
5. Draw blood by making a slight nick in a tail-vein using a scalpel. Collect 25 μl of blood and mix well with 5 μl of 50 mM EDTA. Store the sample at −80 °C (*see* **Notes 8–10**).

3.3 Blood/Urine Gluc Assay

1. Reconstitute CTZ in acidified methanol to 5 mg/ml and store at −20 °C until use (*see* **Note 11**).
2. Prepare a solution of 0.05 mg/ml CTZ by diluting the frozen stock in PBS supplemented with 5 mM NaCl, pH 7.2. Allow this prepared CTZ mixture to stabilize by incubating it for 30 min at room temperature (18–25 °C) in the dark (i.e., wrapped in aluminum foil) (*see* **Note 12**).
3. Transfer 10 μl of each blood sample to a 96-well white or black plate (*see* **Note 13**).
4. Measure Gluc activity using a plate luminometer. Set the luminometer to inject 100 μl/well of the 0.05 mg/ml PBS-diluted CTZ mixture and acquire photon counts for 10 s.

3.4 Bioluminescence Imaging

1. Initialize the IVIS system and allow the temperature of the charge-coupled device (CCD) camera to reach −90 °C.
2. Set the exposure time to utilize the full dynamic range of the CCD detector, while avoiding any pixel saturation (*see* **Note 14**).
3. Place on ice both the methanol-reconstituted 5 mg/ml CTZ solution (i.e., enzyme substrate) and chilled 1× PBS. The solutions will be mixed immediately prior to injection into each animal. Anesthetize the animal (*see* **Note 15**).
4. For each mouse, prepare an injection solution containing 130 μl of PBS and 800 μl/kg of animal body weight of CTZ solution.

Use the insulin syringe to inject the entire volume into the animal via an intravenous retro-orbital injection.

5. Place the animal into the IVIS system and wait 30 s (from the completion of substrate injection) before taking an image of the mouse (*see* **Note 16**).
6. Post-processing and quantification can be performed with Living Image software 3.0, which accompanies the IVIS system. For BLI analysis of primary mammary tumors, photon flux was quantified for each mouse by measuring a circular region of interest encompassing the primary tumor in supine position.

4 Notes

1. 231BR is a sub-line isolated from the brain metastases of a human breast adenocarcinoma cell line. When examined using an intracardiac injection model, numerous systemic metastases—including brain, bone, and lung—are observed.
2. The lentiviral vector carries an expression cassette encoding Gluc and cerulean fluorescent protein (CFP, LV-Gluc-CFP) separated by an internal ribosomal entry site.
3. Cell confluency should be adjusted so that cultures will not become over confluent 2 days after completion of the viral transfection.
4. The addition of a centrifugation step may improve transfection efficiency. After growth media is replaced with lentiviral media, centrifuge the plate at 400–500 × *g* at 4 °C for 1.5–2 h before continuing cell culture.
5. When injecting multiple animals, it is better to prepare aliquots of the cell suspension for use with each animal.
6. In the systemic metastasis model, the blood-based Gluc signal sharply increases 1 day after cell injection, returns to a basal level after a few days, and then increases again as the surviving tumor cells proliferate in the mice [2].
7. Mice are euthanized when animals show clinical symptoms of prolonged distress, signs of neurological impairment, or a loss of more than 20 % body weight; each criterion defines a survival end point.
8. Any body fluid sample (whole blood, plasma, urine, etc.) can be examined with the Gluc assay. Though the urine-based Gluc assay is completely noninvasive, routine collection is demanding and the signal is an order of magnitude lower than that of blood-based Gluc. Alternatively, while plasma-based Gluc assay provides an order of magnitude higher signal than blood-based

Gluc, it requires an additional centrifugation step and twice the blood volume for adequate plasma collection [2]. Thus, whole blood is regarded as the optimal fluid sample for the Gluc assay. All samples used for the Gluc assay contain 20 % (vol/vol) EDTA solution as an anticoagulant.

9. Blood coagulation interferes with the Gluc signal. Thus, it is important to check the collected blood sample for coagulation before storage at −80 °C.
10. Gluc signals measured from thawed samples of frozen stock provide a more uniform result than signals measured from freshly acquired samples.
11. The acidified methanol solution contains a ratio of 5 μl of 12N HCl to 1 ml of methanol.
12. Maintaining this preparation time is critical to attain consistent and repeatable results.
13. It is important to use white or black plates. Translucent plates allow bioluminescence signals to transfer between wells, generating cross-contamination and an inaccurate measurement.
14. If imaging a single mouse, a field-of-view (FOV) of 10 cm × 10 cm is sufficient. If three mice are to be examined, a FOV of 12.5 cm × 12.5 cm should be used. For more than four animals, the FOV should be adjusted to 25 cm × 25 cm.
15. The substrate solution should be stored at a maximum −20 °C (storage at −80 °C is optimal) and kept on ice throughout experiments to prevent precipitation.
16. Since Gluc exhibits flash kinetics, it is critical to keep the time between CTZ injection and image acquisition consistent. It is thus better to image each animal individually than to image groups of animals simultaneously.

Acknowledgements

The authors are grateful for the support of the Steele Laboratory for Tumor Biology (PI: Rakesh K. Jain) and Prof. Bakhos A. Tannous at the Massachusetts General Hospital. This work was supported by grants from the Institute of Medical System Engineering and the Bio Imaging Research Center at Gwangju Institute of Science and Technology (GIST); Basic Research Projects in High-tech Industrial Technology at GIST; the Institute of Medical System Engineering at GIST; the Bio & Medical Technology Development Program and Basic Science Program through the National Research Foundation of Korea Grants 2011-0019633 and 2012R1A1A1012853, and the Tosteson postdoctoral fellowship at the Massachusetts General Hospital (to E.C.)

References

1. Contag CH (2007) In vivo pathology: seeing with molecular specificity and cellular resolution in the living body. Annu Rev Pathol 2: 277–305
2. Chung E, Yamashita H, Au P et al (2009) Secreted Gaussia luciferase as a biomarker for monitoring tumor progression and treatment response of systemic metastases. PLoS ONE 4: e8316
3. Wurdinger T, Badr C, Pike L et al (2008) A secreted luciferase for ex vivo monitoring of in vivo processes. Nat Methods 5:171–173
4. Hewett JW, Tannous B, Niland BP et al (2007) Mutant torsinA interferes with protein processing through the secretory pathway in DYT1 dystonia cells. Proc Natl Acad Sci U S A 104: 7271–7276
5. Badr CE, Hewett JW, Breakefield XO, Tannous BA (2007) A highly sensitive assay for monitoring the secretory pathway and ER stress. PLoS ONE 2:e571
6. Bao R, Connolly DC, Murphy M et al (2002) Activation of cancer-specific gene expression by the survivin promoter. J Natl Cancer Inst 94:522–528
7. Bao R, Selvakumaran M, Hamilton TC (2000) Use of a surrogate marker (human secreted alkaline phosphatase) to monitor in vivo tumor growth and anticancer drug efficacy in ovarian cancer xenografts. Gynecol Oncol 78: 373–379
8. Hiramatsu N, Kasai A, Hayakawa K et al (2006) Secreted protein-based reporter systems for monitoring inflammatory events: critical interference by endoplasmic reticulum stress. J Immunol Methods 315:202–207
9. Meng Y, Kasai A, Hiramatsu N et al (2005) Continuous, noninvasive monitoring of local microscopic inflammation using a genetically engineered cell-based biosensor. Lab Invest 85:1429–1439
10. Kim HJ, Kim YH, Lee DS et al (2008) In vivo imaging of functional targeting of miR-221 in papillary thyroid carcinoma. J Nucl Med 49: 1686–1693
11. Kodack DP, Chung E, Yamashita H et al (2012) Combined targeting of HER2 and VEGFR2 for effective treatment of HER2-amplified breast cancer brain metastases. Proc Natl Acad Sci U S A 109:E3119–E3127
12. Tannous BA (2009) Gaussia luciferase reporter assay for monitoring biological processes in culture and in vivo. Nat Protoc 4:582–591
13. Badr CE, Tannous BA (2011) Bioluminescence imaging: progress and applications. Trends Biotechnol 29:624–633
14. Tannous BA, Kim DE, Fernandez JL et al (2005) Codon-optimized Gaussia luciferase cDNA for mammalian gene expression in culture and in vivo. Mol Ther 11:435–443
15. Korpal M, Yan J, Lu X et al (2009) Imaging transforming growth factor-beta signaling dynamics and therapeutic response in breast cancer bone metastasis. Nat Med 15:960–966
16. Bhaumik S, Gambhir SS (2002) Optical imaging of Renilla luciferase reporter gene expression in living mice. Proc Natl Acad Sci U S A 99:377–382
17. Palmieri D, Bronder JL, Herring JM et al (2007) Her-2 overexpression increases the metastatic outgrowth of breast cancer cells in the brain. Cancer Res 67:4190–4198

Chapter 13

Bioluminescence Imaging of Fungal Biofilm Development in Live Animals

Greetje Vande Velde, Soňa Kucharíková, Patrick Van Dijck, and Uwe Himmelreich

Abstract

Fungal biofilms formed on various types of medical implants represent a major problem for hospitalized patients. These biofilms and related infections are usually difficult to treat because of their resistance to the classical antifungal drugs. Animal models are indispensable for investigating host–pathogen interactions and for identifying new antifungal targets related to biofilm development. A limited number of animal models is available that can be used for testing novel antifungal drugs in vivo against *C. albicans*, one of the most common pathogens causing fungal biofilms. Fungal load in biofilms in these models is traditionally analyzed postmortem, requiring host sacrifice and enumeration of microorganisms from individual biofilms in order to evaluate the amount of colony forming units and the efficacy of antifungal treatment. Bioluminescence imaging (BLI) made compatible with small animal models for in vivo biofilm formation is a valuable noninvasive tool to follow-up biofilm development and its treatment longitudinally, reducing the number of animals needed for such studies. Due to the nondestructive and noninvasive nature of BLI, the imaging procedure can be repeated in the same animal, allowing follow-up of the biofilm growth in vivo without removing the implanted device or detaching the biofilm from its substrate. The method described here introduces BLI of *C. albicans* biofilm formation in vivo on subcutaneously implanted catheters in mice. One of the main challenges to overcome for BLI of fungi is the hampered intracellular substrate delivery through the fungal cell wall, which is managed by using extracellularly located *Gaussia luciferase*. Although detecting a quantifiable in vivo BLI signal from biofilms formed on the inside of implanted catheters is challenging, BLI proved to be a practical tool in the study of fungal biofilms. This method describing the use of BLI for in vivo follow-up of device-related fungal biofilm formation has the potential for efficient in vivo screening for interesting genes of the pathogen and the host involved in *C. albicans* biofilm formation as well as for testing novel antifungal therapies.

Key words *Candida albicans*, Biofilm, Bioluminescence imaging, *Gaussia* luciferase, Catheter, Animal model, Mouse, Coelenterazine, Infectious diseases, Noninvasive, In vivo

1 Introduction

In the past decades, the emergence of fungal infections caused by opportunistic pathogenic yeasts (e.g., *Candida albicans*, *Cryptococcus neoformans*) and molds (e.g., *Aspergillus fumigatus*)

Christian E. Badr (ed.), *Bioluminescent Imaging: Methods and Protocols*, Methods in Molecular Biology, vol. 1098, DOI 10.1007/978-1-62703-718-1_13, © Springer Science+Business Media New York 2014

has increased significantly and has become a common challenge in clinical practice, especially among patients who are immunocompromised or hospitalized with serious underlying diseases [1–3]. Increased numbers of fungal diseases are related to the increasing number of severely ill or immune-compromised patients and in particular to the growing use of implanted medical devices in patients [2, 4, 5]. *Candida albicans* is an opportunistic human fungal pathogen responsible for the majority of yeast infections in these patients [3, 6–8], causing infections that can range from superficial to systemic and life-threatening candidiasis [7].

Candida biofilms and its characteristics have been mostly elucidated in vitro [9]. However, these models cannot account for numerous host and infection-site variables which are important during the development of biofilm-related infection in humans. Recently, in vivo central venous catheter (CVC) *C. albicans* biofilm models have been introduced in rats [10], rabbits [11], and mice [12]. These models are technically demanding and of very low throughput. To address this issue, a rat in vivo subcutaneous foreign body infection model has been adapted to study *C. albicans* biofilms [13, 14]. This subcutaneous catheter rat model system was validated for testing antifungal drugs against mature *C. albicans* biofilms [15]. Although this model cannot account for factors such as shear stress and nutrient flow, the suitability of the model to evaluate the efficacy of antifungal drugs on biofilm development was shown, allowing more rapid screening under in vivo conditions [15].

In these biofilm models, fungal load is traditionally analyzed *post mortem* by enumeration of the number of colony forming units attached to the catheter fragments and from individual host organs to determine the extent and kinetics of dissemination during infection. Only one time point per animal can be measured and therefore large groups of animals are needed. These end-point assays do not always provide a full understanding of the dynamic steps in *Candida* biofilm development, nor do they account for interindividual variation. Noninvasive imaging has the potential to deliver data over the biologically relevant time scale from individual animals in vivo, obviating the need to sacrifice animals at different stages of biofilm development. Among other imaging techniques, bioluminescence imaging (BLI) has emerged as a powerful method to analyze infectious diseases in animal models and microbial viability in particular [16]. Although BLI is well established for imaging different (microbial) cell populations, for yeast cells and other fungal pathogens this is particularly challenging. Because of their cell wall, the diffusion of the necessary substrate for the light producing reaction is limited, hampering detection of a BLI signal from deeper infection sites in vivo [17]. To overcome this problem, a synthetic *C. albicans* codon-optimized version of the gene for the naturally secreted *Gaussia princeps* luciferase (*gLUC*) can be used by tagging *gLUC* to the *C. albicans PGA59* gene, which encodes a glycosylphosphatidylinositol (GPI)-linked cell wall protein [18]. As the luciferase will be present extracellularly, problems with the intracellular delivery of the substrate (coelenterazine, CTZ) are avoided. This expression system has made BLI of superficial *C. albicans* infections possible [18].

Here we describe a multi-temporal noninvasive imaging assay to follow biofilm formation under in vivo conditions in mice and rats, avoiding the need to sacrifice the animals which results in statistical, ethical, economic, and experimental advantages. The use of in vivo imaging reduced the time, number of animals and costs for the analysis of biofilm models and increased the versatility of the current animal models to the use of different transgenic mouse strains. We typically image the animals at baseline (day 0, after catheter implantation), on day 2 when a mature biofilm is formed, and on day 6 when a mature biofilm is sustained. The here described imaging protocol will be readily translatable for use with other in vivo biofilm models such as the CVC model, for the assessment of antifungal therapies and the validation in transgenic models to study microbial and host factors influencing biofilm formation.

2 Materials

2.1 Candida Albicans Strains

1. *Candida albicans* SC5314 (wild-type), a clinical isolate [19]. The genome of this clinical isolate was successfully sequenced.
2. *C. albicans* SKCA23 strain (named SKCA23-*ACTgLUC*) expressing *gLUC* fused to the endogenous *PGA59* gene under the control of *ACT1* (actin) promoter (this promoter is active in the yeast as well as hyphal stage of fungal growth). To construct this strain, *C. albicans* SC5314 was transformed with the Clp10::*ACT1p-gLUC59* plasmid [18], which was integrated in the *RPS10* locus, in order to constitutively express *C. albicans* codon-optimized *Gaussia princeps* luciferase gene (*gLUC*) at the cell wall. The proper integration was verified by PCR analyses.

2.2 Cell Culture and Media

1. Standard cell culture medium: RPMI-1640, with L-glutamine and without sodium carbonate buffered with MOPS, pH 7.0.
2. 1× Phosphate buffered saline (for 1 L of 10× PBS: 80 g NaCl, 2 g KCl, 14.4 g Na_2HPO_4, 2.4 g KH_2PO_4).
3. Sterile water.
4. YPD agar plates (1 % yeast extract, 2 % bacto-peptone, 15 % agar, supplemented with 2 % glucose).
5. Fetal bovine serum (FBS).
6. Cell culture plates: 96-well and 24-well black polystyrene plates (Greiner Bio-One, Cellstar®).

7. Catheters: polyurethane (*see* **Note 1**) triple-lumen intravenous catheters (2.4 mm diameter, Certofix Trio S730, BBraun, *see* **Note 2**), the triple-lumen part cut into 1 cm long fragments.
8. Incubator at 37 °C incubator for static incubation of YPD plates.

2.3 Animals

Before you start planning animal experiments, seek approval from your local bioethics committee and follow the respective guidelines for animal handling. Always take care that you carry out all aspects of animal experiments in compliance with national and supranational regulations regarding animal handling and welfare.

1. Mice: female Balb/C mice 8 weeks of age, kept in individually ventilated filter top cages with free access to standard food and water ad libitum (*see* **Note 3**).
2. Rats: Sprague-Dawley female rats of 200 g (Janvier, France), kept in individually ventilated filter top cages with free access to standard food and water ad libitum.
3. Immune suppression: suppress the immune system of the animals (*see* **Note 4**) 24 h before catheter implant and continue during the entire experiment by adding dexamethasone (0.4 mg/l; Fagron SAS, France) to their drinking water.
4. During immune suppression, also add antibiotics (e.g., ampicillin sodium powder 0.5 g/l) to the drinking water to avoid possible bacterial infections.

2.4 Anesthesia and Surgery

1. Anesthesia for mice: prepare 1 ml of anesthetic cocktail by mixing 75 μl of ketamine (Ketamine1000®, 100 mg/ml) with 100 μl of medetomidine (Domitor®, 1 mg/ml) and 825 μl of sterile saline. Administer intraperitoneally (i.p.) 60–80 μl of anesthetic cocktail per 10 g body weight, resulting in a dose of 45–60 mg/kg ketamine and 0.6–0.8 mg/kg medetomidine. For reversal of anesthesia, dilute 50 μl atipamezole (Antisedan®, 5 mg/ml) in 4.95 ml saline, administer i.p. 100 μl per 10 g body weight as antidote, resulting in a dose of 0.5 mg/kg.
2. Anesthesia for rats: prepare 1 ml of anesthetic cocktail by mixing 600 μl of ketamine with 400 μl of medetomidine. Administer i.p. 100 μl of anesthetic cocktail per 100 g body weight, resulting in a dose of 60 mg/kg ketamine and 0.4 mg/kg medetomidine. For reversal of anesthesia, dilute 1 ml atipamezole in 4 ml saline; administer i.p. 100 μl per 100 g body weight as antidote, resulting in a dose of 1 mg/kg.
3. Xylocaine gel (2 %) (AstraZeneca) as local anesthetic for the skin.
4. Terramycin/polymyxin-B ophthalmic ointment (Pfizer), to apply on the eyes to prevent them from drying/infecting while the animal is under anesthesia.

5. Electric trimmer/razor or disposable razors for shaving the back of the animals.
6. Clean, sterile surgical tools: fine scissors, tweezers, surgical thread, needle, and tools for suturing the skin.
7. Sterile gazes, iodo-isopropanol (1 %) as a disinfectant for the skin.
8. Heating plate or mat to keep the animals warm while they are under anesthesia.
9. Extra clean cages kept warm (on a warm plate or with a lamp) to keep animals separate until they are fully recovered from anesthesia and surgery.

2.5 BLI

1. BLI camera: We have used an IVIS 100 system (Caliper LS, Perkin Elmer, USA) and Living Image software (version 2.50.1 for PC, provided by the manufacturer) for measuring and quantifying luminescence in vitro (*see* **Note 5**) and in vivo (*see* **Note 6**).
2. Coelenterazine solutions: native coelenterazine (CTZ) was purchased from Nanolight Technologies (USA) and stored at −80 °C according to the manufacturer's instructions. 5 mg/ml CTZ were dissolved in acidified ethanol according to the manufacturer's instructions. This stock solution can be stored at −80 °C for the duration of the experiments, typically 6–9 days. For in vitro BLI, 6 μM CTZ working solution was prepared by diluting the stock solution 1:2,000 in sterile PBS. For in vivo BLI, we prepared a 1.2 mM working solution by diluting the stock solution 1:10 in sterile PBS (*see* **Note** 7). Foresee 100 μl CTZ working solution per catheter trio. Insulin syringes (0.3 ml) were used for injections. CTZ is light-sensitive, always keep it in the dark (e.g., cover recipients with aluminium foil).

3 Methods

Follow national regulations for handling microbial pathogens class 2. Perform all handlings with open *C. albicans* suspensions in a vertical laminar air flow cabinet (LAF cabinet) to avoid contamination of cell cultures.

3.1 In Vitro Biofilm Formation

1. Grow the *C. albicans* strains you want to include in your experiment (e.g., *gLUC*-positive strain and wild-type as a control) for 8–12 h (e.g., overnight) on YPD plates at 37 °C. Afterwards, take a loop of *C. albicans* cells; dissolve in 1 ml of 1× PBS and wash twice with 1 ml of PBS. Count the cells (e.g., with a Bürker chamber) to prepare *C. albicans* cell suspensions in RPMI-1640 medium at an appropriate cell density (as mentioned below) for subsequent use in the specific experiment.

2. Inoculate a sterile black (*see* **Note 8**) 96-well cell culture plate by adding 100 μl per well of a 1×10^7 cells/ml *C. albicans* cell suspension in RPMI-1640. Allow the *C. albicans* cells to adhere to the bottom of the wells by incubating the plate for 90 min at 37 °C. Wash the wells twice with PBS and then add 200 μl of fresh RPMI-1640 medium to each well. Incubate the plate again at 37 °C for the duration of the experiment to allow mature biofilm formation on the bottom of the wells (*C. albicans* forms mature biofilms within 24 h–48 h).
3. To grow biofilms inside polyurethane catheters, pre-incubate 1 cm catheter pieces overnight with FBS at 37 °C. Place each catheter fragment in a clean 24-well plate (1 catheter piece per well) and add 1 ml of a *C. albicans* cell suspension (5×10^4 cells/ml). Make sure that catheters are submerged in the medium and not floating on the top. Further incubate the plate for 90 min at 37 °C to allow cell attachment onto the substrate. Wash the catheters twice with 1 ml PBS, transfer them to a clean plate, submerge them in fresh RPMI-1640 medium, and incubate at 37 °C for the duration of the experiment.

3.2 BLI of In Vitro Biofilm Formation

1. Freshly prepare CTZ working solution (*see* Subheading 2) for in vitro experiments and initialize the BLI camera (*see* **Note 9**).
2. Take the culture plates containing biofilms (formed on the bottom of 96- or 24-well polystyrene cell culture plates or biofilms formed inside catheter pieces) out of the incubator.
3. Pipette the culture medium off and wash the biofilms twice with 1 ml of sterile PBS (*see* **Note 10**), end by taking out all the PBS from the wells.
4. Insert the plate in the camera, set the appropriate FOV and find the right imaging position for the plate by taking photographs. Set the settings for acquisition of the BLI image. Typically, we acquired BLI images with a 5–20 s exposure time depending on the signal, at medium binning.
5. Add swiftly but carefully (pipette under a 45° angle against the well wall) the CTZ working solution to the wells (100 μl/well for 96-well plates, 500 μl/well for 24-well plates), preferably with a multichannel pipette (*see* **Note 11**).
6. Immediately thereafter, insert the plates in the BLI-camera and start acquisition of consecutive frames until maximum signal is reached (*see* **Note 12**). Save the acquisitions you need.
7. Quantify the BLI signal using the processing software of your BLI camera and report as photon flux per second (p/s) for a given region of interest (ROI) with a fixed size, covering the well or catheter (Fig. 1 panels a, b).

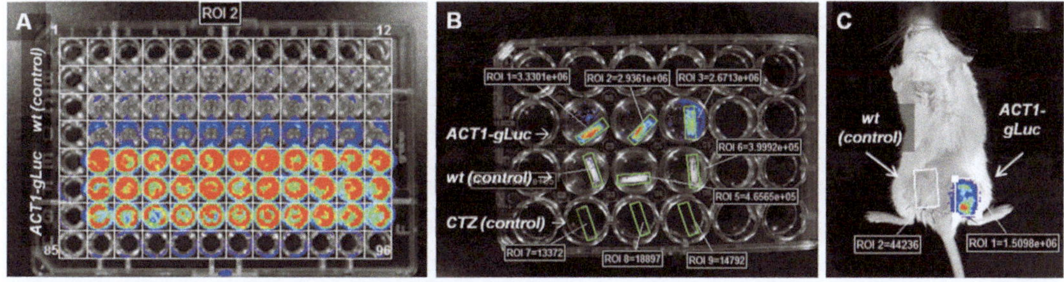

Fig. 1 Regions of interest (ROIs) for BLI signal analysis. Panel **a**: for the analysis of the BLI signal from biofilms formed on the bottom of wells of a 96-well culture plate, an 8 × 12 grid can be used to quantify the photon flux from each well separately. Panel **b**: a rectangular ROI is placed over the catheter piece. Panel **c**: a rectangular ROI is placed over each catheter trio

3.3 In Vivo BLI of Biofilm Formation on Implanted Catheters

1. Incubate the catheter pieces in FBS (100 %) at 37 °C overnight.
2. Place the serum-coated catheters into separate microcentrifuge tubes and add 1 ml of the *C. albicans* cell suspension (5×10^4 cells/ml prepared in RPMI-1640).
3. Incubate for 90 min at 37 °C for adhesion of the cells to the catheters.
4. After the adhesion period, place catheters on ice as fast as possible.
5. Wash the catheters twice with 1 ml PBS, transfer them individually into clean, sterile microcentrifuge tubes, and place on ice for transport to the surgery room for implantation.
6. Prepare the animal for surgery. Anesthetize the mouse by injection (i.p.) of ketamine/Domitor anesthetic cocktail, wait a few minutes until the animal is fully asleep (check by pinching the tail or a paw: the animal should not react to that). Apply ophthalmic ointment on the eyes, shave the lower back of the animal, apply some xylocaine gel on the skin for local anesthesia of the area where you will implant the catheters (wait for a minute to work), disinfect the skin with iodine–isopropanol.
7. Make a small incision in the skin (approximately 0.5 cm–1 cm), carefully dissect the subcutis with a scissor to create two (for mice) or three (for rats) subcutaneous tunnels, make each tunnel about 1.5 cm long and 1 cm wide (Fig. 2).
8. Insert three catheter pieces (inoculated with the same *C. albicans* strain) in each tunnel, taking care that they fit next to each other in a horizontal arrangement and that they do not over each other. This way, you can implant six (for mice) or nine (for rats) catheter fragments in total and test biofilms formed by different *C. albicans* strains in the same animal, e.g., SC5314-gLUC-positive and wild-type (control).

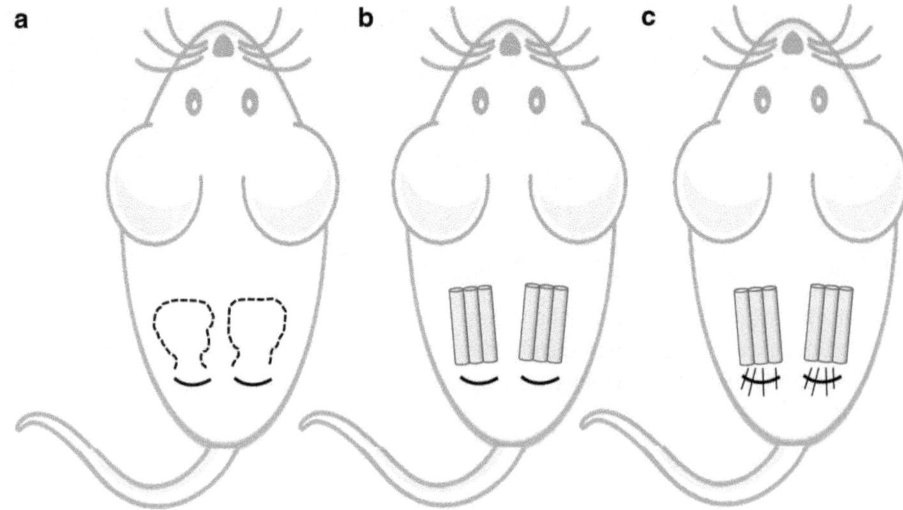

Fig. 2 Catheter implantation. Start by making two incisions in the skin (*full black lines* in panel **a**). Next, by inserting scissors under the skin through these incisions, dissect the subcutaneous space to create two tunnels of about 1.5 cm long and 1 cm wide (*dashed lines* in panel **a**). Make sure that the tunnels do not make contact with each other. Through the incision, insert three catheters next to each other in each subcutaneous space (panel **b**). Close the incisions in the skin by separate surgical sutures (panel **c**). Pay attention that you place the sutures well under the catheters, not above

9. Close the incisions by sutures and disinfect the wounds.
10. Wake the animal up by administering Antisedan i.p. to reverse anesthesia (*see* **Note 13**).
11. In vivo BLI: Prepare fresh in vivo CTZ working solution just before the start of the BLI session and initialize the BLI camera (*see* **Note 9**).
12. Anesthetize the animals using an induction box (comes usually with the BLI camera). Use a gas mixture of isoflurane in oxygen, N_2O/O_2 or air, at 2–3 % for mice and 3–4 % for rats. Rats are best anesthetized and imaged one by one, mice can be put asleep by maximum of 4 at a time in the induction box and then sequentially imaged.
13. After induction, maintain anesthesia in the induction box and in the imaging chamber at 1.5–2 % for mice and 2–3 % for rats, at a flow rate of the carrier gas of 0.5 l/min.
14. Before starting the imaging session, put the imaging plate in position A, which corresponds to a FOV of 10 cm. Reassure yourself of the right position of the anesthesia outlets and animal by placing a sleeping animal in the box and taking some photographs until it is in the correct imaging position, in the FOV right under the camera.

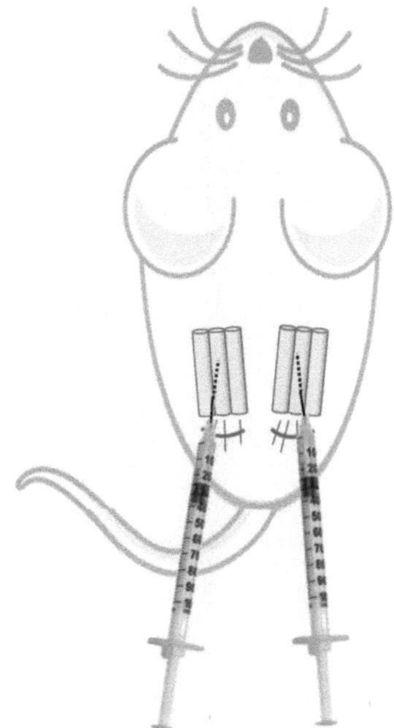

Fig. 3 CTZ injection for in vivo BLI. Insert the syringes in such a way that the largest part of the needle is subcutaneously (*dashed lines*) to avoid leakage of CTZ out of the needle entrance point. Apply the CTZ on top of the catheter trio (end of *dashed line*) simultaneously

15. Prepare two syringes containing each 100 μl of the CTZ solution. Take one animal and use a nose cone with gas anesthesia to keep it asleep on the bench during the injection. Bring the needles of the syringes one by one in place subcutaneously above the catheters as depicted in Fig. 3 (*see* **Notes 14–16**). Use both hands to inject the CTZ simultaneously on top of the catheters (*see* **Note 17**).
16. Immediately after injection, place the animal on the warm plate in the camera box and start the image acquisition. A typical BLI image is depicted in Fig. 4. Acquire consecutive scans with acquisition times ranging from 20 to 60 s (depending on the signal intensity) until the maximum signal intensity is reached (*see* **Note 18**). During the acquisition of the next frame, you can measure the BLI signal of the previously acquired frames by placing a ROI over each catheter trio and measuring the signal intensity (Fig. 1 panel c).
17. Repeat from **step 13** for the next animal(s).
18. Analyze the BLI data and report the BLI signal intensity of each catheter trio as photon flux per second for a rectangular ROI of fixed size placed over each catheter trio.

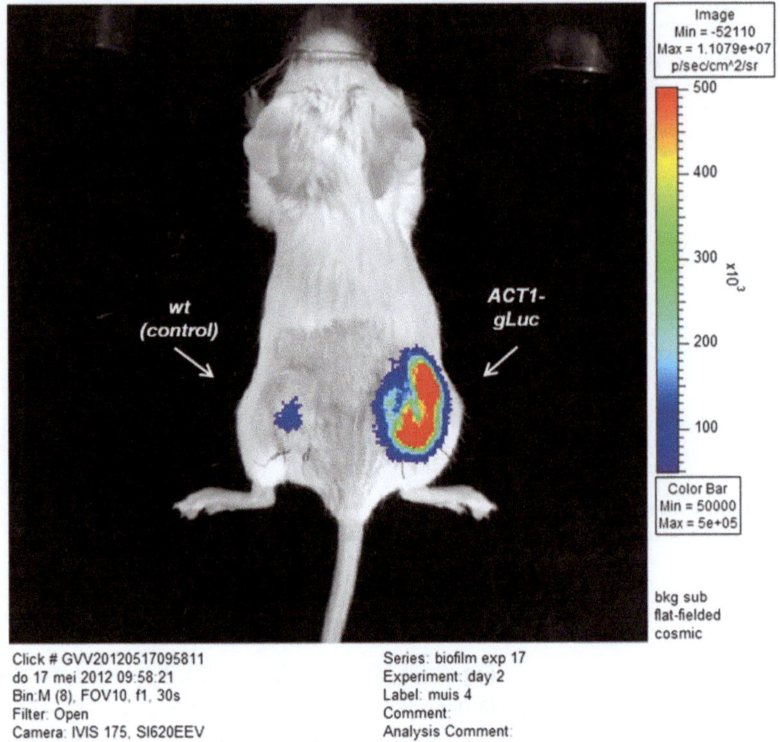

Fig. 4 Typical BLI image from a 2-day-old mature biofilm formed inside 3 catheters implanted left and right under the skin of a mouse. *Left*: biofilms formed by wild-type *C. albicans* SC5314, right biofilms formed by *C. albicans* SC5314 expressing ACT-gLUC

3.4 Explantation of Catheters for Ex Vivo Biofilm Evaluation

After the final imaging time point, recover the catheter pieces from the animals for further ex vivo analysis of the biofilms.

1. Prepare microcentrifuge tubes with 1 ml PBS (each catheter will go into a separate microcentrifuge tube) and label them beforehand.
2. Euthanize the animals by cervical dislocation.
3. Disinfect the skin of the back with 0.5 % chlorhexidine in 70 % alcohol.
4. Make two new incisions above the catheters and remove the catheter fragments one by one from under the subcutaneous tissue using sterile tweezers (*see* **Note 19**).
5. Wash the catheter twice with 1 ml sterile PBS and put it in the microcentrifuge tube (prepared in **step 1**).
6. Keep the tubes with catheters on ice or in the fridge for short storage before further processing (i.e., quantification of biomass or with microscopy techniques).

7. Fungal burden quantification: Wash the cell culture plates, in vitro evaluated catheters, or explanted catheters twice with 1 ml PBS, sonicate for 10 min at 40,000 Hz in a water bath sonicator (Branson 2210), and vigorously vortex for 30 s. Make 1:10 and 1:100 dilutions and plate 100 μl of the original samples, 1:10 and 1:100 dilutions on YPD agar plates in duplicate. Incubate the plates for 2 days at 37 °C, after which you can count colony forming units (CFUs). Recalculate to fungal load per fragments (CFUs/catheter piece) taking into account the dilution factor and the plated volume.
8. Microscopical evaluation: To examine the three-dimensional structure of the biofilms formed on the inside of the explanted catheters, you can consider performing fluorescence microscopy, confocal laser scanning microscopy, or scanning electron microscopy. For these procedures, we refer to [14].

4 Notes

1. Different materials may have different properties for attachment of fungal cells and subsequent biofilm development.
2. Intravenous catheters are often made radiopaque for ease of detection with X-ray once they are brought into place in the body. The substances added to make catheters radiopaque can elicit phosphorescent properties. We found the majority of catheters commonly used in hospitals to be strongly phosphorescent, which confounds BLI results. Therefore, in our study we used Certofix catheter from B. Braun because it is not phosphorescent. If you would use another type of catheter, carefully check for phosphorescence by imaging it in the BLI camera at regular intervals, and check for an increase in signal after exposing the catheters to light, e.g., by opening the scanner door and resuming BLI scanning.
3. We use Balb/C mice as this is a common background strain for many transgenic mouse models. For BLI, it is important that they are white (or nude) as the pigmentation of the skin, even when shaving the animals, absorbs a substantial amount of the BLI signal.
4. The biofilms will also develop without immune suppression of the host, but the fungal load on the catheters may vary substantially. When suppressing the immune system of both rats and mice, the fungal load recovered from the catheters is much more consistent (see also [15]), reducing the standard deviations and hence reducing the amount of animals needed to reach a given power in your experiment. Therefore, we recommend you to use the protocol for immune suppression before you start the catheter implantation.

5. For measuring luminescence in vitro, you can also use any luminometer or essentially any plate reader that is suitable for measuring luminescence. Some can even be equipped with injectors for controlled addition of CTZ. The biggest advantage of placing the plates in the BLI system is that a luminometer/plate reader will not provide an image of your plate.
6. The IVIS system is delivered with a very comprehensive manual, it is worth reading through (the first part of) the manual and you will be able to get started with the camera and the software easily.
7. When diluting the stock solution to make the in vivo CTZ working solution, a small precipitate will form. According to the manufacturer this should not happen, but in our hands using the prescribed dilution protocol, we could never avoid having some extent of precipitation. As far as we know, other users experience the same problem. It is imperative to dilute the CTZ immediately prior to injection as this will limit the precipitation problem.
8. For BLI measurements, it is recommended to use black (or white) cell culture plates, with or without transparent bottom. When not using black plates, there will be BLI signal scattering from adjacent wells into the well you are measuring in case you have high signal intensities. In case you do not have a black plate at hand, you can also use a transparent plate and poor black drawing ink in between wells to reduce this "cross talk." After your BLI experiment is finished, you can use a pipette to take out the ink for later reuse. This is not ideal as there will still be some "cross talk" where the wells are touching each other. This "cross talk" will have some influence on the quantification of your BLI results.
9. Starting up BLI camera: Before you start imaging, you need to login and initialize the system (button: "initialize"). If the IVIS system is only used sporadically, it can be kept off in between imaging sessions. In that case, it will need to cool down before you can start imaging which can take 2–4 h. In most cases, however, the system is left on to keep the CCD camera constantly cooled. Initializing the system takes only few minutes and you will be able to start almost immediately with your imaging experiment.
10. Especially when you are dealing with biofilms formed on the bottom of wells, take care not to detach the biofilm from the bottom of the plates while washing.
11. When doing in vitro BLI measurements, you could consider making fresh CTZ and then incubate it in the dark for ~30 min before you add it to the wells. This way, the difference in

auto-oxidation of the coelenterazine solution will be minimal in case you cannot add CTZ to all the wells simultaneously. When you set up your experiments, always take non-luminescent wild-type controls and control wells where you only add CTZ along in your analysis for comparison with unspecific background signal. On the other hand, if you are interested in measuring the signal kinetics make sure you add CTZ "as simultaneously as you can" (when pipetting with a multichannel pipette, make sure that in every pipetting move you take along experiment and controls) and start image acquisition as fast as you can after CTZ addition.

12. In practice, you will almost immediately reach the maximum signal when performing in vitro BLI measurements as diffusion of CTZ into the biofilms is fast, after which the signal will steadily decrease. Depending on your research question, you can take about three frames at maximum signal with the settings of your choice or keep acquiring short frames during a certain period of time if you are interested in signal kinetics.

13. We do not recommend keeping the animal under ketamine/Domitor anesthesia and proceed to BLI imaging immediately after surgery while it is still anesthetized. We recommend letting the animal completely recover from the injection anesthesia before starting isoflurane anesthesia for the imaging session. The latter procedure was well tolerated by the animals.

14. We evaluated signal kinetics in vivo upon different administration routes of the CTZ (intravenous, intraperitoneal, subcutaneous) and compared the BLI signal. We found that CTZ does not have a favorable biodistribution upon iv or ip injection. For this reason, we recommend to apply CTZ topically.

15. Upon CTZ injection, it is crucial that no CTZ leaks out of the little hole you make in the skin upon inserting the syringe. That is why it is crucial to keep ~1 cm of the needle under the skin when you inject, as shown in Fig. 3. CTZ shows some extent of auto-oxidation. Therefore, any "leaked" CTZ on the skin of the animal would give rise to unspecific BLI signal that would significantly influence the BLI measurement.

16. Injecting CTZ subcutaneously is obvious at the first time point, as there is a lot of subcutaneous space because of the tunnels created during surgery. However, this will heal over time and there will be fibrous tissue formed around the catheters, making injecting subcutaneously less evident, especially when doing BLI at later time points (6–9 days after catheter implantation). Take care not to inject *intra*-cutaneously. In this case, you will notice immediately by a "bobble" appearing due to the intradermal CTZ injection. No or only a very low CTZ signal will be seen.

17. In case experiments are performed on rats that have three (or more) tunnels with catheter trios, ask a colleague to assist you with injecting CTZ in all the tunnels simultaneously.
18. At the first time point, the maximum signal is reached in less than 1 min after CTZ administration. The more later after catheter implantation surgery, the longer it takes to reach this maximum signal, presumably because of the increase of fibrous tissue around the catheters that makes the CTZ diffuse more slowly towards the biofilms.
19. Make sure to decontaminate the tweezers in between catheter pieces having biofilms from different *C. albicans* strains.

Acknowledgements

This work was funded by KU Leuven PF "IMIR," the FWO Research community on biology and ecology of bacterial and fungal biofilms (FWO: WO.026.11N), KU Leuven PDMK 11/089 fellowship and FWO postdoctoral fellowship to SK. We thank Christophe d'Enfert for providing us with the Clp10::ACT1p-gLUC59 plasmid.

References

1. Warnock DW (2006) Fungal diseases: an evolving public health challenge. Med Mycol 44(8):697–705
2. Nucci M, Marr KA (2005) Emerging fungal diseases. Clin Infect Dis 41(4):521–526
3. Pfaller MA, Diekema DJ (2010) Epidemiology of invasive mycoses in North America. Crit Rev Microbiol 36(1):1–53
4. Eggimann P, Garbino J, Pittet D (2003) Management of *Candida* species infections in critically ill patients. Lancet Infect Dis 3(12):772–785
5. Eggimann P, Garbino J, Pittet D (2003) Epidemiology of *Candida* species infections in critically ill non-immunosuppressed patients. Lancet Infect Dis 3(11):685–702
6. Pemán J, Salavert M (2012) Epidemiología general de la enfermedad fúngica invasora. Enferm Infecc Microbiol Clin 30(2):90–98
7. Wisplinghoff H, Bischoff T, Tallent SM, Seifert H, Wenzel RP, Edmond MB (2004) Nosocomial bloodstream infections in US hospitals: analysis of 24,179 cases from a prospective nationwide surveillance study. Clin Infect Dis 39(3):309–317
8. Enoch DA, Ludlam HA, Brown NM (2006) Invasive fungal infections: a review of epidemiology and management options. J Med Microbiol 55(Pt 7):809–818
9. Nett J, Andes D (2006) *Candida albicans* biofilm development, modeling a host-pathogen interaction. Curr Opin Microbiol 9(4):340–345
10. Andes D, Nett J, Oschel P, Albrecht R, Marchillo K, Pitula A (2004) Development and characterization of an in vivo central venous catheter *Candida albicans* biofilm model. Infect Immun 72(10):6023–6031
11. Schinabeck MK, Long LA, Hossain MA, Chandra J, Mukherjee PK, Mohamed S, Ghannoum MA (2004) Rabbit model of *Candida albicans* biofilm infection: liposomal amphotericin B antifungal lock therapy. Antimicrob Agents Chemother 48(5):1727–1732
12. Lazzell AL, Chaturvedi AK, Pierce CG, Prasad D, Uppuluri P, Lopez-Ribot JL (2009) Treatment and prevention of *Candida albicans* biofilms with caspofungin in a novel central venous catheter murine model of candidiasis. J Antimicrob Chemother 64(3):567–570
13. Van Wijngaerden E, Peetermans WE, Vandersmissen J, Van Lierde S, Bobbaers H, Van Eldere J (1999) Foreign body infection: a new rat model for prophylaxis and treatment. J Antimicrob Chemother 44(5):669–674
14. Řičicová M, Kucharíková S, Tournu H, Hendrix J, Bujdáková H, Van Eldere J, Lagrou K, Van Dijck P (2009) *Candida albicans*

biofilm formaton in a new in vivo rat model. Microbiology 156(Pt3):909–919

15. Kucharíková S, Tournu H, Holtappels M, Van Dijck P, Lagrou K (2010) In vivo efficacy of anidulafungin against mature *Candida albicans* biofilms in a novel rat model of catheter-associated Candidiasis. Antimicrob Agents Chemother 54(10):4474–4475
16. Hutchens M, Luker GD (2007) Applications of bioluminescence imaging to the study of infectious diseases. Cell Microbiol 9(10): 2315–2322
17. Doyle TC, Nawotka KA, Kawahara CB, Francis KP, Contag PR (2006) Visualizing fungal infections in living mice using bioluminescent pathogenic *Candida albicans* strains transformed with the firefly luciferase gene. Microb Pathog 40(2):82–90
18. Enjalbert B, Rachini A, Vediyappan G, Pietrella D, Spaccapelo R, Vecchiarelli A, Brown AJ, d'Enfert C (2009) A multifunctional, synthetic *Gaussia princeps* luciferase reporter for live imaging of *Candida albicans* infections. Infect Immun 77(11):4847–4858
19. Gillum AM, Tsay EY, Kirsch DR (1984) Isolation of the *Candida albicans* gene for orotidine-5'-phosphate decarboxylase by complementation of *S. cerevisiae* ura3 and *E. coli* pyrF mutations. Mol Gen Genet 198(1): 179–182

Chapter 14

Bioluminescent Imaging of Bacteria During Mouse Infection

Jonathan M. Warawa and Matthew B. Lawrenz

Abstract

Diagnostic imaging is a powerful tool that has recently been applied towards the study of infectious diseases. Optical imaging of bioluminescently labeled bacteria in infected animals allows for real-time analysis of bacterial proliferation and dissemination during infection without sacrificing the animal. Imaging also allows for tracking of disease progression in an individual subject over time, has the potential to reveal previously overlooked sites of infection, and reduces the number of research animals used in pathogenesis studies. Here, we describe the use of a deep-cooled CCD camera imager to record light emitted from bacteria during infection. We also describe the process of correlating bioluminescence to bacterial numbers by ex vivo imaging of necropsied tissues. Together these techniques can be used to estimate bacterial burdens in host tissues both in vivo and ex vivo using bioluminescent imaging.

Key words Optical diagnostic imaging, Bioluminescence, Surrogate disease model, IVIS Spectrum, Alternative approach

1 Introduction

Conventional techniques to study microbial pathogenesis in mammalian models have primarily relied on end point assays (e.g., survival analysis or lethal dose 50 [LD_{50}]) or on sacrificing animals at designated time points to characterize bacterial dissemination to specific tissues. The development of noninvasive approaches to image disease progression, without having to sacrifice the animal, allows for a temporal, real-time analysis of host–pathogen interactions. The use of such noninvasive techniques also facilitates reduction of animal numbers required for these studies—one of the three R's of good laboratory animal practice [1].

In this chapter, we describe the use of optical diagnostic imaging to characterize bacterial disease progression within a small animal host. For optical imaging, animals are infected with bacteria that emit light or photons. Currently, two light-emitting modalities exist: fluorescent or bioluminescent bioreporters.

Christian E. Badr (ed.), *Bioluminescent Imaging: Methods and Protocols*, Methods in Molecular Biology, vol. 1098,
DOI 10.1007/978-1-62703-718-1_14,

Bioluminescent bioreporters have been favored because they tend to be more sensitive than equivalent fluorescent bioreporters and demonstrate a dynamic relationship to bacterial viability, which is lacking in fluorescent systems because of the long half-life of fluorescent proteins in inactivated bacteria [2]. Since first described by Contag et al. [3], bioluminescent bioreporters using bacterial *luxCDABE* luciferase operons have been developed for many different pathogens. A major advantage of the bacterial luciferase systems over eukaryotic luciferase systems is that the bacterial *lux* operon also synthesizes the substrate for the luciferase enzyme [2]. Therefore, bioreporters using a bacterial system do not require addition of exogenous substrate (usually through injection prior to imaging) to facilitate optical imaging. This also eliminates the potential complication of bioavailability issues with an exogenously added substrate. Furthermore, bacterial and eukaryotic luciferases have different spectral emissions, allowing for the potential to use both in the same system (i.e., monitor bacteria with a bacterial luciferase bioreporter and immune response with a eukaryotic luciferase bioreporter).

Bioluminescent bioreporters for specific pathogens may be available from commercial sources, requests from other laboratories, or designed and engineered in advance in house. Technical details on the development of specific bioreporters will not be discussed here, but several important criteria should be considered when selecting or designing an appropriate system. First, the reporter should be as sensitive as possible without altering the growth of the bacterium. This can be achieved by increasing the copy number of the reporter (i.e., expression from a plasmid vs. chromosome) or by engineering a highly active promoter to drive the expression of the *luxCDABE* operon. Second, the reporter must be maintained by the bacterium throughout the entire course of the animal infection. Integration of the *luxCDABE* reporter into the genome or onto a plasmid with a toxin–antitoxin selection system can improve stability of the reporter genes. Finally, the expression of the reporter should be constitutive during infection so that bioluminescence closely correlates with the number of viable bacteria regardless of the time or location being imaged. This can also be controlled through empirical selection of a promoter to control the transcription of the luciferase operon. However, as each new model is being developed, extensive in vivo and ex vivo characterization of the bioreporter in regard to bioluminescence and actual bacterial numbers needs to be determined to establish if bioluminescence directly correlates to bacterial load.

The following technique focuses on using optical imaging and subsequent data analysis to characterize bacterial infection in small animal models using intranasal instillation of *Yersinia pestis* as an example. However, similar approaches can be applied to a variety of bacterial pathogens and routes of inoculation.

2 Materials

2.1 Bacterial Infection

1. *Yersinia pestis* LuxP_{tolC} (Fully virulent *Y. pestis* CO92 bioluminescent bioreporter strain [4]).
2. Spectrophotometer.
3. P20 pipettor.

2.2 Whole Animal Live Imaging

1. Appropriate mouse strain (*see* **Note 1**).
2. IVIS Spectrum (Perkin Elmer).
3. Living Image 3.0 Software (Perkin Elmer) for image collection from the IVIS Spectrum and performing post-capture data analysis.
4. XGI-8 anesthesia system and isoflurane/oxygen inhalation anesthetic to maintain sedation during image capture.
5. f/AIR isoflurane scavenging charcoal filters: filters should be weighed routinely, and retired when they have absorbed 50 g of additional weight.
6. Anesthesia manifold with nose cones and dividers, transparency, and black construction paper (*see* **Note 2**).
7. Forceps.

2.3 Ex Vivo Tissue Imaging

1. Sterile black 24-well plates (*see* **Note 3**).
2. Sterile surgical instruments.
3. Sterile Whirl-Pak 1 oz bags (Nasco): the sealing strip removed and the dry weight of the bag recorded in advance.
4. Sterile 1× phosphate buffered saline (PBS).
5. Pathogen-specific nutrient agar plates.

3 Methods

3.1 Animal Husbandry and Preparing the IVIS Spectrum

1. One week prior to infection studies, mice are transitioned onto an alfalfa-free rodent chow to reduce the alfalfa-related luminescent background in the gut.
2. Three to four days prior to infection studies, mice are shaved using a veterinary clipper with a size 30–40 blade. Removal of fur increases the sensitivity of detection of bioluminescent bacteria (*see* **Note 1**).
3. Individual mice within a group are identified using black Sharpie tail banding, toe tattooing, or ear tags (*see* **Note 4**).
4. The day before imaging begins, launch the Living Image software and initialize the IVIS Spectrum. Ensure that the Auto Background Settings/Dark Charge measurements are current

Table 1
Dilution schedule for *Y. pestis* inoculums

Tube	1	2	3	4	5	6	7
Dilution (in 1 ml PBS)	1 OD_{600}	1:3	1:10	1:10 inoculum	1:10 plate for CFU	1:10 plate for CFU	1:10 plate for CFU

by manually acquiring the Background (under the Acquisition tab) (*see* **Note 5**).

5. Prepare the sample stage within the IVIS Spectrum. Place a transparency on the sample stage and then one sheet of black construction paper on top of the transparency. Insert the nose cones and dividers into the anesthesia manifold on the imaging stage (*see* **Note 2**).

3.2 Preparation of Y. Pestis Inoculum and Animal Inoculation

1. Three days before mouse inoculation, inoculate a BHI agar plate with a scraping from −80 °C glycerol stock of frozen culture of *Y. pestis* LuxP_{tolC}. Grow for 2 days at 26 °C.
2. The day before mouse inoculation, inoculate a 5 ml BHI broth culture with a 1 in. streak of multiple colonies from the BHI plate. Grow for 6–8 h on a roller drum at 26 °C.
3. Determine the OD_{600} of the 26 °C culture using the spectrophotometer and dilute bacteria to an OD_{600} of 0.05 in 10 ml of prewarmed BHI + 2.5 mM $CaCl_2$ (in a 125 ml flask).
4. Grow for 16–18 h at 37 °C shaking at 250 RPM.
5. Determine the OD_{600} of the 37 °C culture using the spectrophotometer and perform serial dilutions as shown in Table 1 in 1 ml of sterile 1× PBS. Spot plate 20 μl from dilution tubes 5, 6, and 7 on BHI agar and grow for 2 days at 26 °C to determine the inoculum concentration (*see* **Note 6**).
6. Anesthetize the first group of five mice (*see* **Note 7**) and deliver 10 μl of bacterial suspension from Tube 4 (Table 1) using a P20 pipet to one nare of each mouse. When finished with the fifth mouse, repeat a second 10 μl delivery into the same nare. Return mice to cages and continue with subsequent groups.

3.3 In Vivo Whole Animal Live Imaging

For each model, imaging sessions should be scheduled as appropriate. For acute infections, typically two imaging sessions per day provides sufficient data to observe in vivo disease progression. Prior to inoculation with the bioreporter pathogen, a baseline image of the animals should be acquired to establish background luminescence.

1. Place the first group of mice (5 mice max per group) into an anesthesia induction chamber.

2. Set the XGI-8 anesthesia system to deliver 2 % isoflurane and at a flow rate of 2.5–3 liters per minute (LPM) to the induction chamber (*see* **Note 8**).
3. Allow the mice to reach surgical plane sedation (approximately 5–10 min) so that they remain sedated for transfer to and placement in the IVIS Spectrum. To facilitate efficient transfer, arrange animals within the induction chamber in the order that they will be placed into the IVIS Spectrum.
4. Using the IVIS Acquisition Control Panel, initialize the IVIS spectrum.
5. Adjust the IVIS Acquisition Settings to acquire a luminescent image of five mice (Imaging Mode = Luminescence, Field of View D). Also set the appropriate exposure time. Set the first exposure time to 5 min (Exposure Time = 5 min) (*see* **Note 9**).
6. Turn on anesthesia to the IVIS Spectrum at a flow rate of 0.5 LPM (*see* **Note 8**).
7. Transfer anesthetized mice to the IVIS spectrum, placing their heads into the nose cones, dorsal side up. Use forceps to help center the animals between the dividers and position the animal's legs and tails.
8. Close the door to the IVIS Spectrum and select Acquire on the IVIS Acquisition Control Panel.
9. At the completion of the first image capture of a session, the user is offered the Auto-Save function, allowing for instant saving of all captured images. Click Yes and select or create a folder to save images from this session (*see* **Note 10**).
10. If imaging more than one group, place the next group of mice into anesthesia induction chamber while imaging the first group.
11. (Optional) Once imaging of the dorsal side is completed, flip mice and position each mouse to image ventral side.
12. (Optional) Repeat **step 8** to image the ventral perspective.
13. When imaging is completed for each group, remove animals from the IVIS Spectrum and return animals to their cages. Mice are monitored until they recover from anesthesia, which takes approximately 2 min. for healthy animals. However, recovery times can increase as animals progress to later stages of disease.
14. Repeat imaging with the remainder of the groups of mice (repeat **steps 7–13** as necessary).
15. At the completion of imaging, the IVIS is disinfected by disposal of the liners (black construction paper and transparency), soaking of nose cones in disinfectant, and surface decontamination of the IVIS imaging stage (*see* **Note 11**).

3.4 Ex Vivo Tissue Imaging

At designated time points or at the end of an imaging experiment, infected animals can be sacrificed and specific tissues harvested to determine bioluminescence ex vivo. These analyses also allow the researcher to correlate the bioluminescent signal observed to actual bacterial numbers.

1. After completing the last in vivo imaging session for a specific group, humanely euthanize animals using an AVAMA approved technique.
2. Using aseptic technique, remove tissues from one mouse and place each tissue into individual wells of a 24-well black plate. If harvested, transfer 100 μl of host fluids (e.g., bronchoalveolar lavage [BAL] or blood) to separate wells. Because tissues from several hosts may be imaged in a single 24-well plate, ensure that each plate has at least one empty well to serve as a bioluminescence blank.
3. Place the lid on the 24-well plate and transfer to the IVIS imaging stage.
4. Adjust the IVIS Acquisition Settings to acquire a luminescent image of the 24-well plate. Also set the appropriate exposure time. Set the first exposure time to 1 min (Imaging Mode = Luminescence, Field of View = C, Exposure Time = 1 min) (*see* **Note 9**).
5. Close the door to the IVIS Spectrum and select Acquire on the IVIS Acquisition Control Panel.
6. When imaging is completed, remove plate from IVIS Spectrum. Aseptically transfer the tissues into labeled, pre-weighed Whirl-Pak bags. Seal bags and store on ice as necessary. Proceed with necropsy of the next mouse and ex vivo tissue imaging.
7. After tissues from all mice have been imaged, determine the tissue weight [Tissue Weight = (weight of Whirl-Pak + tissue) − (weight of Whirl-Pak)]. Add 500 μl of sterile 1× PBS to the Whirl-Pak and re-seal. Macerate tissues within the Whirl-Pak by rolling over the tissue with a 10 ml serological pipette.
8. Make serial dilutions of macerated tissues in sterile 1× PBS and spot plate 20 μl of each dilution on an agar plate.
9. Incubate agar plates and enumerate CFU of bacteria for each tissue (*see* **Note 12**).

3.5 Image Analysis

One of the strengths of live animal optical imaging is the ability to observe the progression of the infection in the same animal over the course of the disease (temporal analysis). Once all images over the course on the experiment have been acquired, the progression or dissemination of the infection can be analyzed by quantifying light detected from regions of interest from the mouse at each time point. By analyzing the same region for each time point, changes in bacterial numbers (as a function of bioluminescence) can be calculated over time. Furthermore, optical imaging can be used to identify virulence defects of mutant bacteria or efficacy of therapeutics by comparing similar regions from different mice at the same time point.

1. Combine sequential images of the same group into a series by clicking the Browse button from the menu bar. Select desired images from each time point and click OK.
2. Highlight sequential images from the Living Image Browser window and click Load as Group.
3. To quantitate bioluminescence, Regions of Interest (ROI) need to be designated for each animal/tissue. To add ROIs to individual images, click the Display All button.
   (a) Select the first image to add an ROI (*see* **Note 13**). Select the desired ROI shape from the ROI Tools menu (add only 1 shape). Adjust the size and position of the ROI to include the region that you want to analyze (*see* **Note 14**).
   (b) Generate copies of first ROI (right click on ROI) for each animal in the same image.
   (c) Position each ROI in the desired location but do not change the size of the ROI (i.e., all ROIs for a specific location/tissue will be the same size).
   (d) Once all ROIs are drawn for the first image, right click on the image and select Copy All ROIs. Right click on the other images from the series and paste ROIs. Position each ROI in the desired location but do not change the size of the ROI (i.e., all ROIs for a specific location/tissue will be the same size).
   (e) Save images and ROIs will be added to the Sequence View (Fig. 1).
4. From the Sequence View, uncheck the Individual box. This will allow a single color scale to be used for all images in the sequence (*see* **Note 15**). Check the Apply to All box and change the Units to Photons. To view the bioluminescent signal from the ROIs, select ROI measurements from the View menu.
5. Highlight ROIs you wish to analyze (or select all) and export to Excel.
6. To compare ROIs between time points or different mice, use the Avg. Radiance ($p/s^2/cm^2/sr$) data. Normalize this data by subtracting the Avg. radiance from the uninfected time point (*see* **Note 16**).

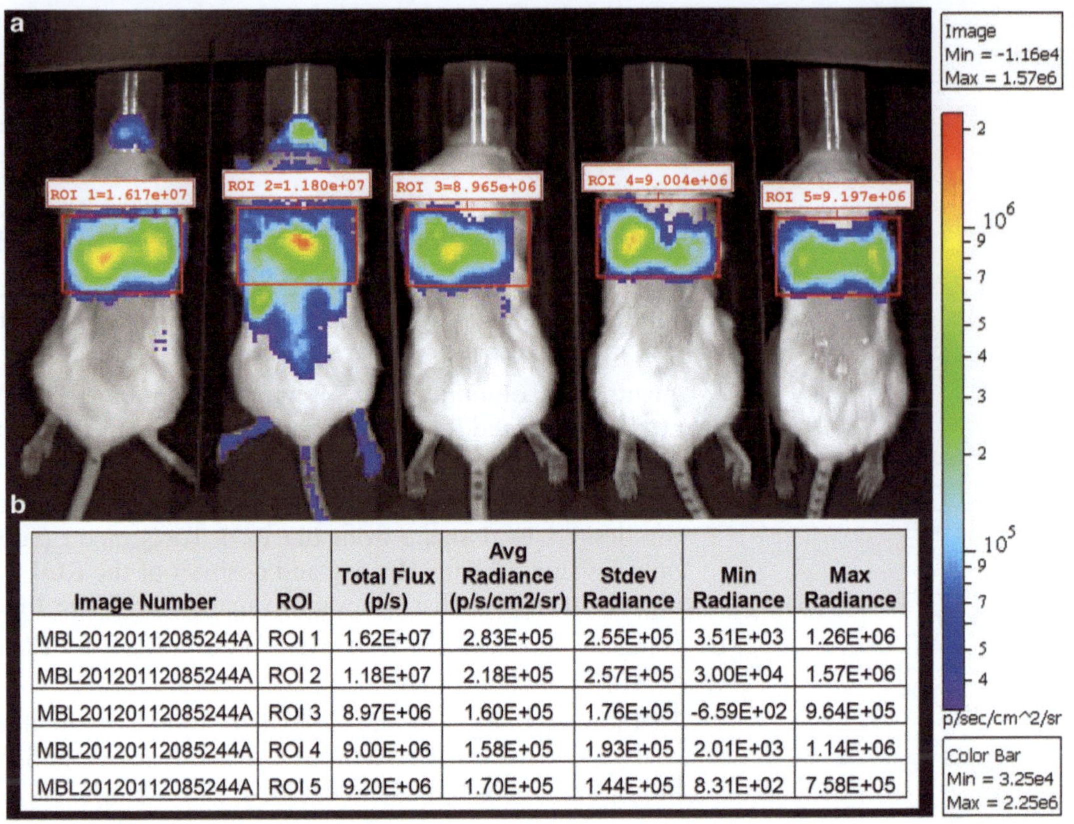

Image Number	ROI	Total Flux (p/s)	Avg Radiance (p/s/cm2/sr)	Stdev Radiance	Min Radiance	Max Radiance
MBL20120112085244A	ROI 1	1.62E+07	2.83E+05	2.55E+05	3.51E+03	1.26E+06
MBL20120112085244A	ROI 2	1.18E+07	2.18E+05	2.57E+05	3.00E+04	1.57E+06
MBL20120112085244A	ROI 3	8.97E+06	1.60E+05	1.76E+05	-6.59E+02	9.64E+05
MBL20120112085244A	ROI 4	9.00E+06	1.58E+05	1.93E+05	2.01E+03	1.14E+06
MBL20120112085244A	ROI 5	9.20E+06	1.70E+05	1.44E+05	8.31E+02	7.58E+05

Fig. 1 Defining ROIs for infected animals. (**a**) To determine the amount of bioluminescence emitted from a specific region of infected mice, square ROIs (shown in *red*) were drawn and positioned over the thoracic cavity of each mouse. Total flux for each ROI is shown in the white ROI tags. (**b**) Measurement window for the ROIs in (**a**)

3.6 Correlating Bacterial Numbers to Bioluminescence

It is possible to define a correlation coefficient to estimate bacterial burden in ex vivo host tissues from a rapid bioluminescence measurement. The following analysis bridges bacterial enumeration approaches with direct bioluminescent imaging of necropsied tissue to derive a correlation coefficient for each host tissue infected with a bioluminescent pathogenic bacterial strain. The key benefit of this approach is facilitating rapid future analysis of bacterial burdens in host tissue, leaving the tissue intact for other analyses.

1. Bacterial enumeration studies conducted in Subheading 3.4 are used to provide bacterial burden measurements for each infected host tissue. Host tissues have bacterial burdens presented as CFU/tissue. For host fluids such as BAL and blood, bacterial burdens are presented as CFU/ml.

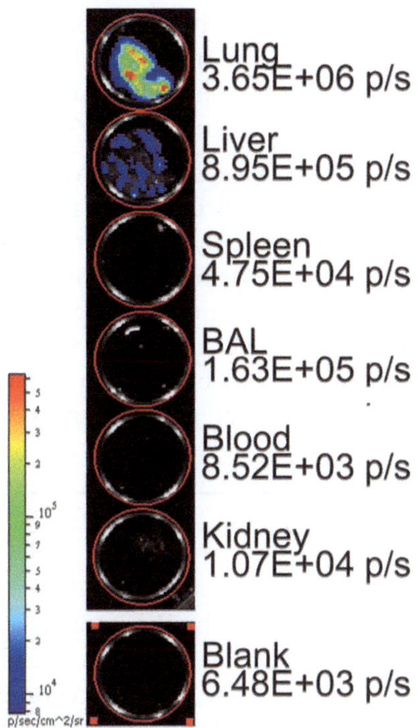

Fig. 2 Measured bioluminescence from host tissues ex vivo. Mice were infected with luminescent bacteria in a respiratory tract infection and necropsied at moribund disease. Infected host samples were collected into a black 24-well plate and an IVIS Spectrum was used to measure the bioluminescence. Identical round ROIs were fit to each well (*red circle* depicted), and total flux (p/s) measurements were exported to Excel as a measure of bioluminescence for each tissue/host fluid

2. Bioluminescence measurements obtained at Subheading 3.4, **step 5** are further analyzed as follows:
   (a) Using the Living Image software, an ROI is fit to the captured image as described in Subheading 3.5. Grid ROIs are available for fitting standard sized microtiter plates, and a 4 × 6 grid is sized to a 24-well plate to capture ROIs for all wells. Alternatively, identically sized ROIs may be copied and pasted onto sample-containing wells, as discussed in Subheading 3.5, **step 3** (Fig. 2).
   (b) With units set to Photons, view ROI measurements and export to Excel.
   (c) Use the Total Flux (p/s) values to calculate the total bioluminescence present in each well, using an empty well to provide a measure of background which is subtracted from all tissue bioluminescence measurements. Host tissue measurements are stated as p/s/tissue for measurement of the entire tissue.

(d) If 100 μl of host fluid (e.g., BAL or blood) was measured in a well, the background is subtracted, and the total bioluminescence (p/s) is multiplied by 10 to state data in units of p/s/ml.

3. Graphing software is used to plot bioluminescence as a function of bacterial burden for both host tissues (p/s/tissue vs. CFU/tissue) and host fluids (p/s/ml vs. CFU/ml). GraphPad Prism may be used to perform log transformations of both the bioluminescence and bacterial burden and define a best fit linear regression relationship. The equation of this correlation can be used in future studies to perform a rapid estimate of bacterial burden in necropsied host tissues.

4 Notes

1. The ability to detect low numbers of bioluminescent bacteria within the mouse can be significantly impacted by the coat color of the mouse. The use of albino mice will increase sensitivity and is recommended. The animal's fur will also decrease the sensitivity of the bioreporter and it is highly recommended that the fur be removed prior to imaging. Shaving is preferred to the use of topical depilatory creams as the regrowth of hair is slower in shaved mice. Hairless mice are also available but tend to be associated with immunodeficiencies and may not represent suitable host models.
2. Work with animals infected with bacterial pathogens should be conducted within biosafety cabinets (BSC) as per Institutional Biosafety Committee (IBC) guidelines. Risk assessments should be conducted when working with pathogens which can be shed in high titer in feces or urine, or those that are communicable by aerosol transmission. Biocontainment chambers are commercially available (e.g., Perkin Elmer XIC-3) for high-risk organisms which require biocontainment during the imaging process [5]. Risk assessments should be conducted with your IBC to determine whether infected animals may be safely transported to an IVIS for imaging outside of a biocontainment imaging chamber.
3. Direct imaging of infected host tissues followed by bacterial enumeration requires a sterile black 24-well plate, which may not be commercially available as a sterile product. We order non-sterile black 24-well plates (Greiner Bio-One, Product # 662174) and clear lids (Greiner Bio-One, Product # 656190) and then assemble and sterilize lidded plates using ethylene oxide sterilization.

4. We have observed luminescent signal from colored Sharpie (red and blue) or tattoo inks (green) and recommend the use of black inks only. If Sharpie tail banding is used for identification, the bands will need to be renewed every other day.
5. A full measurement of background/dark charge for all factory defined imaging conditions will take >2 h, thus it is important to power on the IVIS and perform these measurements well in advance of an image capture. The IVIS should remain powered on during the entire imaging study to allow for daily updates to the dark charge measurements. The software retains up to 48 h of dark charge measurements, and requires these measurements to ensure the ability to compare IVIS data collected on different days and different machines.
6. The CFU/ml/OD_{600} correlation coefficient of culture may vary between laboratories and strains of bacteria. Therefore it is recommended that CFU/ml/OD_{600} and appropriate dilution schedule be determined empirically for each user and strain prior to the imaging study.
7. Mice can be anesthetized by different methods. However, we have found that for *Y. pestis* anesthetization with ketamine–xylazine as opposed to isoflurane results in a more consistent and reproducible instillation of bacteria into the lungs.
8. The dosing schedule for isoflurane should be empirically determined for the mouse line being used. If a single anesthesia delivery system is being used (e.g., XGI-8), the induction and maintenance concentrations of isoflurane being delivered are identical and it is therefore not possible to use a higher dose of isoflurane for initial anesthesia induction, and a second concentration for maintenance if staging groups through the IVIS. We have empirically determined that 2 % isoflurane balances induction and maintenance for both BALB/c and C57BL/6J mice, but sufficient time should be given to ensure surgical plane sedation during induction, such that mice do not revive during transfer to the IVIS. In certain cases, a second independent induction chamber with its own gas mixer may be advisable.
9. We use a 5 min exposure as a starting point for imaging animals and 1 min for tissues. However, depending on the progression of the infection and the bioreporter being used, the CCD detector may become saturated with this exposure time. If this occurs, the software will report an error message notifying the user that a region is saturated. If the user wants to use the luminescent data as a quantitative measurement of bacterial numbers, the animals/tissues will need to be re-imaged with a lower exposure time to achieve a non-saturated image. Imaging beyond 5 min exposures does not justifiably increase sensitivity.

Additional advanced capture settings may be employed to optimize image capture, including binning and F-stop settings.

10. While it is not required to Auto-Save, it is highly recommended to reduce accidental loss of data. It is also possible to access this function by selecting the folder you want your images saved to in the Acquisition tab (Acquisition → Auto-Save). The folder that images are saved to can also be changed in the Acquisition tab throughout the experiment.
11. We have determined that most disinfectants have a high luminescent background which can significantly impact the sensitivity of detecting weakly luminescent bacteria. The most suitable disinfectant for conducting bacterial optical diagnostic imaging is 70 % ethanol, which leaves negligible background luminescence and no long-term residue on the equipment.
12. Bacterial burdens are calculated for each host tissue/fluid by back-calculating the concentration of bacteria present in the original tissue homogenate, then multiplying the concentration (CFU/ml) by the estimated volume of tissue homogenate (ml) to yield total CFU per tissue. The volume of the tissue homogenate is calculated by adding 500 μl of PBS to the estimated tissue volume (measured weight of the tissue (Subheading 3.3, **step** 7) multiplied by a density of 1 g/ml). For host fluids such as BAL and blood, bacterial burdens can be stated in CFU/ml.
13. Typically it is easier to add the first ROI to the last time point of the series. This image will likely have the largest signal area. Once this ROI is drawn, it is copied to the other images in the series, maintaining the same size ROI for all images in the series. This allows for direct comparison of ROI signal over time.
14. If you are analyzing multiple regions/tissues, you can generate multiple ROIs for each mouse. For example, you can generate a square to analyze the thoracic cavity/lungs and an oval to analyze the spleen.
15. To generate images for presentations or publication, the color scale can be adjusted to remove background signal. This adjustment only alters the visual image, it does not change the signal detected in the ROIs.
16. While not as intense as autofluorescence background signal, we have observed slight autoluminescent signal from the mice, especially during longer exposures. By imaging the mice before infection, you can subtract this background signal from image analysis during infection. This may be of particular importance when performing therapeutic studies where detection of sterile immunity is desirable.

Acknowledgement

This work was supported by internal funding at the University of Louisville.

References

1. Food Security Act of 1985. Public Law 99-198 16 U.S.C. 1985. p 3801–3862
2. Meighen EA (1991) Molecular biology of bacterial bioluminescence. Microbiol Rev 55(1): 123–142
3. Contag CH et al (1995) Photonic detection of bacterial pathogens in living hosts. Mol Microbiol 18(4):593–603
4. Sun Y et al (2012) Development of bioluminescent bioreporters for in vitro and in vivo tracking of *Yersinia pestis*. PLoS One 7(10): e47123
5. Alderman TS, Frothingham R, Sempowski GD (2010) Validation of an animal isolation imaging chamber for use in animal biosafety level-3 containment. Appl Biosaf 15(2):62–66

Part IV

Therapeutic Imaging

Chapter 15

Cell-Based Bioluminescence Screening Assays

Romain J. Amante and Christian E. Badr

Abstract

Drug screening is an essential and widely used technique for drug discovery in various biomedical fields notably in oncology. Here we describe a functional screening assay based on the bioluminescence detection of a secreted luciferase for monitoring cell viability of cancer cells in a high-throughput format. This assay allows the screening of large libraries comprising thousands of compounds and the identification of potential anticancer molecules in a rapid, facile, and cost-effective manner.

Key words Screening, Drugs, Bioluminescence, Gaussia luciferase, Glioblastoma

1 Introduction

Identification of molecules with therapeutic benefit has become more accessible due to the availability of large libraries of chemical and natural products. While libraries comprising hundreds of thousands of compounds are now commercially available, drug discovery and development is still a laborious, expensive, and lengthy process. In an ideal setting, it starts with target identification, followed by compound screen eventually identifying a lead compound which will be optimized for efficacy and pharmacokinetics prior to translation into patients [1]. Methods to identify new drugs from compound libraries have become more accessible due to more affordable screening devices, larger libraries and the emerging of noninvasive techniques for faster and robust assay readout, such as bioluminescence imaging (BLI) [2].

Noninvasive molecular imaging of cultured cells and living animals has opened new avenues to help understand fundamental molecular and physiological processes [3]. BLI, in particular, became a widely used laboratory technique allowing the monitoring of different biological processes in immunology [4], oncology [5], virology [6] and neuroscience [7] among many other fields.

Christian E. Badr (ed.), *Bioluminescent Imaging: Methods and Protocols*, Methods in Molecular Biology, vol. 1098,
DOI 10.1007/978-1-62703-718-1_15,

BLI helps expedite the discovery as well as the development and optimization of new drugs [8]. The sensitive and relatively quick readout of photon emission allows screening for thousands of molecules in a labor, time, and cost-effective manner [9]. In fact, BLI has been widely used to screen thousands of compounds mostly in cell-based assays for target identification, gene signaling networks as well as protein–protein interaction [2, 10, 11]. This imaging modality is also suited for multiplexing with other genomic or proteomic assays for a multiparameter profiling of the complex cellular phenotype [12].

Recently, our group has developed a small-molecule drug screening assay based on the *Gaussia luciferase* (Gluc) [2]. When expressed under a constitutively active promoter, Gluc activity is linear with respect to time and cell number [2, 13]. The natural secretion properties of this luciferase and its high sensitivity as well as the time and cost effectiveness of bioluminescence imaging make this reporter an ideal readout for high-throughput screening [14]. Gluc activity can be measured either from aliquots of the conditioned medium for noninvasive longitudinal studies or directly from the luciferase-expressing cells in their medium.

Glioblastoma cells lines used in this screen were genetically engineered to stably express Gluc together with the Cyan Fluorescent Protein (CFP). Both Gluc and CFP are expressed in a lentivirus vector and the latter is used to assess the transduction efficiency. For our assay we used a constitutively active promoter driving Gluc to monitor cell viability. When investigating compounds that could affect gene expression or gene regulation instead of cell viability, the constitutively active promoter can be substituted with the inducible promoter of choice. These reporters typically consist of one or multiple transcription factors binding sites linked to a minimal promoter to drive the expression of a luciferase. Such bioluminescent reporters are often used in gene-targeted drug screening. In one example, a p53-reporter vector was used to identify small molecules affecting the transcriptional activation of this gene [15]. This screen identified compounds that activate p53 transcription.

It is important to note that while Gluc was our reporter of choice for this particular type of small-molecules screening, it could be replaced with other luciferases such as Firefly, Renilla, or Cypridina (also known as Vargula luciferase, Vluc). Luciferases that use different substrates can be combined together and used in multiplexed bioluminescence based assays.

In this chapter we describe all necessary steps needed to design and perform a compound-screening assay aimed to identify potential small molecules that kill glioblastoma cells.

2 Materials

2.1 Cell Culture

1. U87 cells, American Type Culture Collection (ATCC) HTB-14.
2. Complete medium: Dulbecco's Modified Eagle Medium (DMEM), Fetal Calf Serum 10 %, 100 U Penicillin and 0.1 mg/mL Streptomycin (Pen/Strep).
3. Phosphate Buffered Saline (PBS), sterile.
4. 0.05 % Trypsin/Ethylenediaminetetraacetic acid (EDTA) 1×.
5. A 37 °C, CO_2 regulated incubator.

2.2 Lentivirus Vector Transfection

1. 2 M Calcium Chloride ($CaCl_2$).
2. 2.5 mM 4-(2-hydroxyethyl)-1-piperazineethanesulfonic acid (HEPES).
3. 2× HEPES-Buffered Saline (HEBS), pH 7.00–7.05 (*see* **Note 1**).
4. Plasmids for lentivirus packaging:
   (a) pCSCW-Gluc-IRES-CFP: a lentiviral vector encoding Gluc and the cyan fluorescent protein (CFP) under an SV40 promoter and separated by an internal ribosome entry sequence (IRES) [16] (*see* **Note 2**).
   (b) Cytomegalovirus (CMV) ΔR8.91 (>0.5 μg/μL, Lentivirus).
   (c) Vesicular Stomatitis Virus Glycoprotein (VSVG).
5. 15 cm cell culture plates.
6. SW32 Ti Rotor for ultracentrifugation (optional, for concentrating the lentivirus).
7. 293T cells, American Type Culture Collection (ATCC) CRL-11268 (for lentivirus packaging).
8. Reduced Serum Medium, modification of Eagle's Minimum Essential Medium (Opti-MEM) with Pen/Strep.
9. Syringe filters, 0.45 μM.
10. Polybrene (hexadimethrine bromide).

2.3 Compounds Screening

1. 20,000 compounds from CNS-Set™ library (ChemBridge™) [17] (*see* **Note 3**).
2. 384-Well Optical Bottom Plates.
3. Coelenterazine (CTZ), substrate of *Gaussia Luciferase* enzyme.
4. Dimethyl sulfoxide (DMSO).
5. Temozolomide (TMZ).
6. Tris buffer (30 nM), pH 8.0.
7. Triton X-100.

8. Orbitor RS Microplate Mover (Thermo Scientific).
9. Multidrop Combi (Thermo Scientific), for the dispensing of liquids and shaking plates.
10. Luminometer or a plate reader capable of measuring luminescence. In our case we used the Flexstation 3 (Molecular Devices).
11. Momentum Software, for managing and running the Orbitor RS Microplate Mover, Multidrop Combi, and Flexstation 3 in a fully automated manner.
12. Softmax 5.4 software, for bioluminescence data acquisition and analysis.
13. Multichannel pipettes (for manual screen, *see* **Note 4**). A fixed or adjustable 8- or 12-channel pipette can be use to pipet into a 384-well plate. The multichannel pipettes used should cover a range of 1–100 μL.

3 Methods

3.1 Propagation of Glioblastoma Cells

1. U87 cells are grown in a T75 flask.
2. Aspirate cell culture medium.
3. Wash cells once with 5 mL of PBS.
4. Aspirate PBS.
5. Trypsinize cells with 1 mL (5 min, 37 °C), and then add 3 mL of complete medium to stop the reaction.
6. Detach the remaining attached cells from the flash by pipetting up and down. Collect them at the bottom of the flask and transfer them to a 50 mL tube.
7. Centrifuge cells at $400 \times g$ for 5 min.
8. Aspirate medium and resuspend the pellet with 1 mL of complete medium.
9. After counting, plate 1×10^6 cells in a T75 flask.
10. Repeat this routine every 2 days (80 % of confluence).

3.2 Lentivirus Vector Transfection

1. Day 1: Plate 21×10^6 293T cells in a 150×25 mm dish using complete medium.
2. Day 2: Replace medium 2 h before transfection with fresh complete medium.
   (a) Prepare transfection mix:
   Vial 1: 18 μg pCSCW-Gluc-IRES-CFP vector, 18 μg CMVΔR8.91, 12 μg VSVG, 96.0 μL 2 M $CaCl_2$, up to 780 μL 2.5 mM HEPES.
   Vial 2: 780 μL 2× HEBS.

(b) Mix vial 1 dropwise to vial 2. Vortex for 1 min.
(c) Incubate at room temperature for 20 min.
(d) Add dropwise to the 293T cells. Rock plate to distribute mix evenly.

3. Day 3: Wash 1× with plain DMEM. Replace with 13 mL Opti-MEM/Pen/Strep.
4. Day 4: Harvest virus medium. Replace with 13 mL Opti-MEM/Pen/Strep. Store harvested medium at 4 °C.
5. Day 5: Harvest virus medium and combine with the one of day 4. Spin at 450 × *g* for 5 min and filter through 0.45 μm filter.
6. Optional step: Transfer filtered virus medium to SW32 tubes. Ultracentrifugation at 125,000 × *g* for 98 min. Aspirate the supernatant and resuspend in 100 μL of Opti-MEM.
7. Aliquots of the concentrated lentivirus are stored at −80 °C.
8. The lentivirus is titered as transducing units/ml (t.u./ml) using 293T cells after counting the CFP positive cells using either an epifluorescent microscope or flow cytometry. A typical titer obtained using this packaging method is around 10^8 t.u./ml

3.3 Transduction of U87

1. Plate the cells 1 day before transduction (*see* **Note 5**).
2. After thawing the concentrated CSCW-Gluc-IRES-CFP lentivirus on ice, add 2–10 μL to 15 mL of complete medium, preferably in the presence of 10 μg/ml of Polybrene (*see* **Note 6**) and vortex it for few seconds.
3. The medium of the T75 containing the U87 cells with fresh medium containing the lentivirus.
4. Check the transfection efficiency minimum 6–24 h later (*see* **Notes** 7 and **8**, Fig. 1).

3.4 Cell Dispensing in 384-Well Plate (see Fig. 2)

1. Trypsinize and count the cells using a hemocytometer or an automated cell counter.
2. Prepare a cell suspension of 2×10^4 cells/mL.
3. Dispense 100 μL/well of the cell suspension either manually from column 1 to 24 using a 20–200 μL multichannel pipette or the Multidrop Combi, controlled by Momentum Software (*see* **Notes 9** and **10**).

Wait 24 h to allow cell adhesion before adding compounds.

3.5 Statistical Analysis: Z′ Factor (See Note 11)

1. 24 h after the cell dispensing, dispense 0.1 % DMSO v/v in columns 1–12 as a negative control (*see* **Note 12**), and 100 μM TMZ as a positive control, from column 13 to 24 (*see* **Note 13**). This experiment must be done in triplicate.

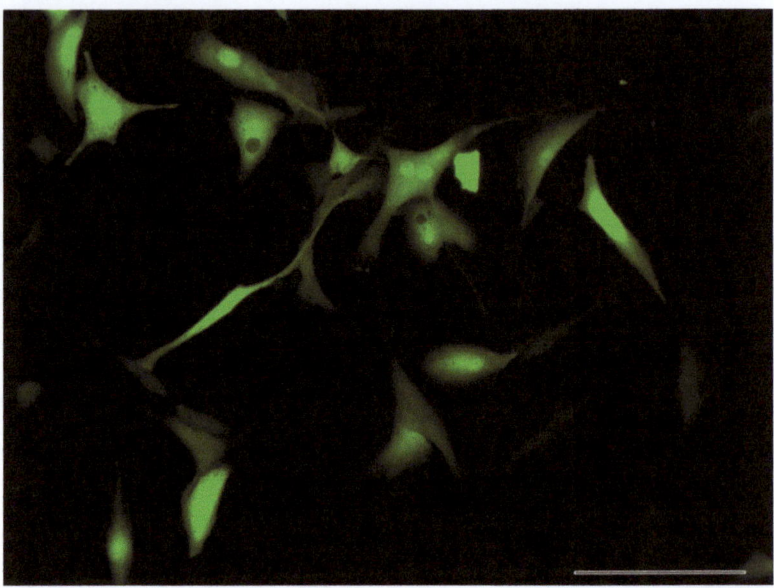

Fig. 1 U87 cells 48 h after transduction with CSCW-Gluc-IRES-CFP (Epifluorescence microscope; Scale 50 μm)

2. After the measurement of the whole plates, it is necessary to determine the Z' factor by using the following formula:

$$Z' = 1 - \left(3 \times \frac{SD_{DMSO} + SD_{TMZ}}{\mu_{DMSO} + \mu_{TMZ}}\right)$$

SD: Standard deviation
μ: Average

3.6 Compound Addition

A fully automated system such as the one described here consists of: an Orbitor RS system, a robotic arm that moves the plates between several devices, a Multidrop Combi for liquid dispensing, and the Flexstation 3 for bioluminescence measurement. Alternatively, the compound addition step might be performed manually using a 1–20 μL multichannel pipette.

1. Prepare a sub-stock (1 mM) of the library (10 mM), by diluting 5 μL in 45 μL of DMSO.
2. Dilute 5 μL of this sub-stock in 45 μL of MEM Medium (100 μM).
3. Add 1 μL of the diluted compounds plate, using the 1–20 μL multichannel pipette or the Multidrop Combi, to the cells in 99 μL of medium (final concentration 1 μM).
4. In each plate, the columns 1 and 24 are reserved for the negative control with 0.1 % DMSO v/v, and positive control with 100 μM TMZ (*see* **Note 14**).

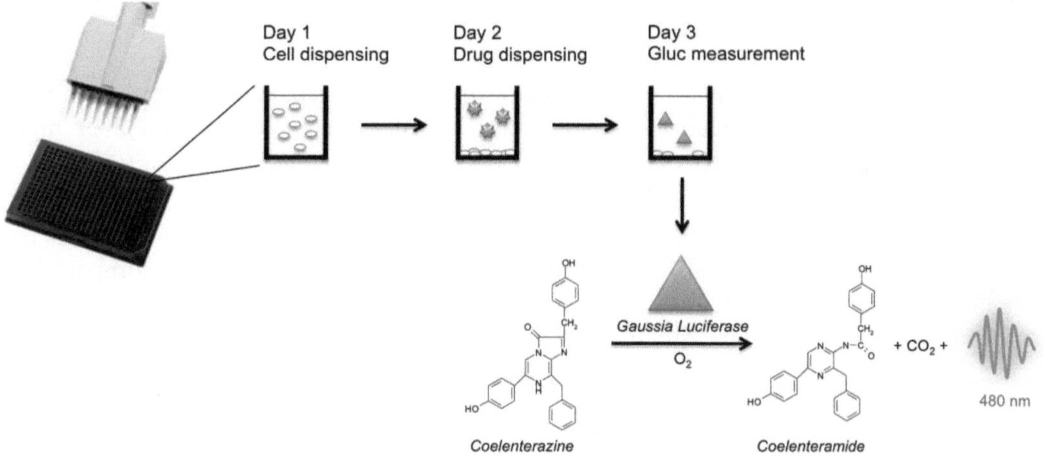

Fig. 2 The chronological steps of the drug screen

5. Gently shake the plates using an orbital shaker for 30 s to ensure equal distribution of the compounds. A quick centrifugation step is recommended when working with small volumes (1–40 μL).
6. Return the plates containing the cells and the added compounds to the incubator.

3.7 Bioluminescence Measurement

The Gluc signal is measured 3 days (72 h) following compound addition (*see* **Note 15**).

1. Prepare 15 μL of coelenterazine substrate/well.
2. Dilute the coelenterazine stock (5 mg/mL) 1/1,250, to obtain a final concentration of 8 μg/mL in Tris buffer, with 0.1 % Triton X-100 (*see* **Notes 16** and **17**).
3. Dispense 15 μL of substrate/well using the 1–20 μL multichannel pipette or the Multidrop Combi.
4. Measure the bioluminescence 1 min after substrate addition, using the Flexstation 3 luminometer with a Softmax Pro Software (*see* **Note 18**), with a 25 ms reading time.

3.8 Hit Validation

1. After performing the screen (typically each compound plate is screened in triplicates), all data can now be plotted and analyzed using Excel spreadsheets.
2. Hits from the primary screen are then identified (*see* **Notes 19** and **20**).
3. Use the selected hits for further validation. This secondary screen will help eliminate any falsely identified hit (*see* **Note 21**).
4. Once the hits from the secondary screen are confirmed, proceed to further testing in dose/time-response experiments and validation on additional cell lines or primary cells.

4 Notes

1. The pH accuracy of this buffer is crucial to achieve high transfection efficiency and should not be lower than 7 or higher than 7.05.
2. We selected the SV40 promoter since it has been shown to be less prone to variation caused by external stimuli such as irradiation, chemotherapeutic drugs etc., as compared to Cytomegalovirus (CMV). The IRES allows the expression of both Gluc and CFP under the same SV40 promoter.
3. The herein described method can be applied to any compound screening library [17].
4. If the screen is to be performed manually, the equipment and software from **steps 5**, **6**, and **8** are not needed.
5. To achieve optimal transduction efficiency, it is important that the cells are no more then 70–80 % confluent on the day of transduction, and 15 h after the last trypsinization.
6. Polybrene is a cationic polymer that increases the transduction efficiency in different cell types.
7. We typically achieve over 90 % transduction efficiency of the lentivirus in U87 cells as assessed by expression of the CFP. Other cell types might be difficult to transduce. If that is the case, it is possible to repeat the transduction step at least 24 h after the initial addition of lentivirus on the cells. Since transduced cells are CFP positive, it is also recommended to sort them by flow cytometry in order to eliminate the non-transduced cells.
8. The transduced cells should be expanded and several cryostocks prepared. This will ensure that all cells used in the screen have the same transduction efficiency and Gluc expression is more homogeneous among the different batches of frozen cells. Moreover, when the screen is performed over several days or weeks, using the frozen cell stocks will provide enough cells for the entire screen, and avoid culturing the cells over a long period of time.
9. It is of utmost importance to plate an adequate number of cells, in order to avoid: (a) having over confluent cells on the day of Gluc measurement (b) reaching bioluminescent signal saturation caused by an excess of Gluc produced in each well of the 384-well plate. We recommend establishing the linearity of Gluc in respect to time and cell number in a 384-well plate format prior to the compound screening. The sensitivity of your assay can greatly vary depending on cell type, lentivirus titer, concentration of luciferase substrate, luminometer, etc.
10. The Multidrop Combi system needs to be primed with cell solution prior to use. An additional volume of 10 mL of cell suspension should be considered for that purpose.

11. The *Z'* factor is a strong indicative of the robustness of the Gluc assay. A proper *Z'* score should show a significant difference between the negative and the positive controls. *Z'* values range between 0 and 1, closer to 1 values are indicative of a robust assay while *Z'* values close to 0 indicate an overlap between positive and negative control thus requiring further optimization of your assay conditions.
12. DMSO is typically the universal solvent in many screening libraries and therefore it is used as a negative control mimicking the final concentration of DMSO (0.1 % DMSO v/v) when compounds are added to the cells.
13. Temozolomide (TMZ) is an alkylating agent clinically used for treatment of GBM patients. Other drugs or compounds that efficiently kill the cell type used in your screen can be used as positive control.
14. Depending on the compound library, empty wells for control addition may vary. Whenever possible, it is advised to scatter the controls throughout the plate and not only on the edges of the plate to avoid false-readings due to positional effect; Outer wells of a 384-well plate can be subjected to liquid evaporation. This is known as the edge effect. Using special high humidity cell culture incubators or low-evaporation plate lids can help minimize this effect.
15. Since Gluc is secreted, it is possible to collect few microliters of conditioned medium (5–15 μL) over multiple periods of time in order to establish time-course curves or to perform the readout from each well in several replicates. The collected aliquots can be stored at −20 °C for several days without any loss of Gluc activity. If using a non-secreted luciferase, an additional step of cell lysis is needed since the active luciferase resides inside of the cells.
16. Coelenterazine is dissolved in acidified methanol (methanol with 1–2 drops of HCl) and a 10 mg/ml stock solution is prepared. Aliquots of coelenterazine are stored at −20 °C and should always be protected from light.
17. Gluc catalyzes a flash-type bioluminescence reaction with a rapid decrease in light emission. It has been previously shown that addition of the detergent Triton X-100 changed Gluc luminescence kinetics and enhanced light stability [18].
18. Due to the flash kinetics of the Gluc reaction, it is ideal to measure each well right after substrate addition. This option might not be available in many plate readers where the reading starts after all wells had been injected with the substrate. Whenever possible, we recommend to use a reader which can inject the substrate and measure the bioluminescence right after. In this configuration, there is no difference due to the timing delay

during dispensing or signal acquisition. During data analysis, it is also possible to mathematically adjust for the signal decay caused by the delay in measurement between the first and the last measured well.

19. When screening for toxic compounds, Gluc levels are expected to be lower than the basal level. In the opposite case, when Gluc values are higher than the basal level, the compound presumably enhances cell proliferation.
20. A selection cut-off for "hit" identification is empirically determined by the screener. For our screen purposes, we set up a threshold of 75 % decrease in Gluc activity. Compounds that cause decrease in Gluc activity greater than 75 % are considered as potential "hits."
21. One disadvantage of reporter genes is the promoter element driving its expression, which is constituted of transcription factor binding sites. These artificial reporters are subject to nonspecific regulation of the screen molecules, which could lead to false or nonspecific readouts. Further some compounds can interfere with Gluc processing through the secretory pathway leading to false positive results. Validation of lead hits using orthogonal viability assays is a prerequisite to ensure that all identified compounds do kill the target cells.

References

1. Willmann JK, van Bruggen N, Dinkelborg LM, Gambhir SS (2008) Molecular imaging in drug development. Nat Rev Drug Discov 7:591–607
2. Badr CE, Wurdinger T, Tannous BA (2011) Functional drug screening assay reveals potential glioma therapeutics. Assay Drug Dev Technol 9:281–289
3. Gaikwad SM, Gunjal L, Junutula AR et al (2013) Non-invasive imaging of phosphoinositide-3-kinase-catalytic-subunit-alpha (PIK3CA) promoter modulation in small animal models. PLoS One 8:e55971
4. McMillin DW, Delmore J, Negri JM et al (2012) Compartment-specific bioluminescence imaging platform for the high-throughput evaluation of antitumor immune function. Blood 119:e131–e138
5. Thompson SM, Callstrom MR, Knudsen BE et al (2013) Molecular bioluminescence imaging as a noninvasive tool for monitoring tumor growth and therapeutic response to MRI-guided laser ablation in a rat model of hepatocellular carcinoma. Invest Radiol 48(6): 413–421
6. Raaben M, Prins H-J, Martens AC et al (2009) Non-invasive imaging of mouse hepatitis coronavirus infection reveals determinants of viral replication and spread in vivo. Cell Microbiol 11:825–841
7. Badr CE, Tannous BA (2011) Bioluminescence imaging: progress and applications. Trends Biotechnol 29:624–633
8. O'Farrell AC, Shnyder SD, Marston G et al (2013) Non-invasive molecular imaging for preclinical cancer therapeutic development. Br J Pharmacol 169(4):719–735
9. Improgo MRD, Johnson CW, Tapper AR, Gardner PD (2011) Bioluminescence-based high-throughput screen identifies pharmacological agents that target neurotransmitter signaling in small cell lung carcinoma. PLoS One 6:e24132
10. Preusser M, Berghoff AS, Capper D et al. (2013) No evidence for BRAF-V600E mutations in gastroeosophageal tumors: results from a high-throughput analysis of 534 cases using a mutation-specific antibody. Appl Immunohistochem Mol Morphol [Epub ahead of print]
11. Choi SM, Kim Y, Shim JS et al (2013) Efficient drug screening and gene correction for treating liver disease using patient-specific stem cells. Hepatology 57(6):2458–2468

12. Feng Y, Mitchison TJ, Bender A et al (2009) Multi-parameter phenotypic profiling: using cellular effects to characterize small-molecule compounds. Nat Rev Drug Discov 8:567–578
13. Badr CE, Wurdinger T, Nilsson J et al (2011) Lanatoside C sensitizes glioblastoma cells to tumor necrosis factor-related apoptosis-inducing ligand and induces an alternative cell death pathway. Neuro Oncol 13:1213–1224
14. Inoue Y, Sheng F, Kiryu S et al (2011) Gaussia luciferase for bioluminescence tumor monitoring in comparison with firefly luciferase. Mol Imaging 10:377–385
15. Wang W, Kim S-H, El-Deiry WS (2006) Small-molecule modulators of p53 family signaling and antitumor effects in p53-deficient human colon tumor xenografts. Proc Natl Acad Sci U S A 103:11003–11008
16. Badr CE, Hewett JW, Breakefield XO, Tannous BA (2007) A highly sensitive assay for monitoring the secretory pathway and ER stress. PLoS One 2:e571
17. ChemBridge | Screening Library | Diversity Libraries. http://www.chembridge.com/screening_libraries/diversity_libraries/index.php#CombiSet
18. Maguire CA, Deliolanis NC, Pike L et al (2009) Gaussia luciferase variant for high-throughput functional screening applications. Anal Chem 81:7102–7106

Chapter 16

Bioluminescence-Based Monitoring of Virus Vector-Mediated Gene Transfer in Mice

Casey A. Maguire

Abstract

In vivo bioluminescence imaging (BLI) is a powerful technology that gives information on biological processes in living animals over multiple time points. Importantly BLI can also yield anatomical localization of signal which can provide important information when performing biodistribution studies of different macromolecules. This is of particular interest for gene therapy vectors such as adeno-associated virus (AAV) vectors in which knowledge of in vivo gene expression profiles help characterize what target tissues or organs the vector may be useful for. It can also be utilized to assess novel vector systems for their ability to overcome specific in vivo barriers of effective gene therapy. Here we describe BLI of AAV-encoded firefly luciferase (Fluc) expression in mice after intravascular delivery. This protocol can be amended for use with different virus vectors (e.g., lentivirus, adenovirus) as well as nonviral gene delivery (e.g., plasmid DNA, liposomes).

Key words In vivo bioluminescence imaging, Firefly luciferase, Virus vector, Adeno-associated virus (AAV), Optical imaging, Mouse models

1 Introduction

In vivo bioluminescence imaging (BLI) has provided in depth knowledge of biological processes such as metastasis [1], transcription factor activation [2], and protein–protein interactions [3]. BLI has become an important tool of gene therapy as it allows a real-time kinetic analysis of gene expression in individual animals [4]. Firefly luciferase (Fluc) is one of the most widely used luciferases for BLI for several reasons. First, the half-life of Fluc is short [5] (2–3 h) allowing accurate measurement of ongoing processes without accumulation effects observed with fluorescent reporters with longer half-lives [6]. Second, there is a relatively inexpensive, water-soluble version of its substrate, D-luciferin, that permits relatively large amounts of substrate to be injected intraperitoneally (i.p.), which is a technically simpler injection route than intravenous (i.v.) injection.

Christian E. Badr (ed.), *Bioluminescent Imaging: Methods and Protocols*, Methods in Molecular Biology, vol. 1098,
DOI 10.1007/978-1-62703-718-1_16,

Finally since Fluc is a cytoplasmic protein, the enzyme and catalyzed light emission are localized to the expression site.

Virus vectors based on adeno-associated virus (AAV) mediate high levels of transgene expression in several preclinical models, including nonhuman primates [7, 8]. In the field of neuroscience, AAV is the most powerful vector tool due to its robust gene transfer to cells of the CNS [9]. Recently several serotypes of AAV have been shown to cross the blood–brain barrier in mice and nonhuman primates, leading to broad-spread gene delivery throughout the brain [10, 11]. Despite the promising results with AAV there are several in vivo biological barriers which hinder the vector for certain applications in the human population. For example, neutralizing antibodies (approximately 30 % of people in North America have neutralizing antibodies to AAV2 [12]) can abrogate gene transfer after i.v. injection. Using BLI to assess novel AAV serotypes or genetically engineered vectors ability to transduce liver in the presence of neutralizing antibodies is a rapid approach to screening multiple vectors. Combined with histological analysis, BLI is an excellent component in the evaluation of the in vivo transduction profiles of virus vectors including AAV.

Here we describe a method for evaluating distribution of transgene expression of AAV-encoded Fluc in mice after tail vein injection.

2 Materials

2.1 Virus Components

1. 293T cells (passage 1–30).
2. Dulbecco's Modified Eagles Medium (DMEM) (high glucose) supplemented with 10 % fetal bovine serum and 1 % Penicillin/ Streptomycin (p/s).
3. Dulbecco's Modified Eagle Medium (high glucose) supplemented with 2 % Fetal bovine serum and 1 % Penicillin Streptomycin.
4. 15 cm diameter tissue culture dishes.
5. Trypsin.
6. Phosphate buffered saline.
7. 2 M $CaCl_2$.
8. 2.5 mM HEPES (4-(2-hydroxyethyl)-1-piperazineethanesulfonic acid) buffer.
9. 2× HEBS (HEPES buffered saline) pH 7.0–7.05

Final concentration: 280 mM NaCl, 50 mM HEPES, 1.5 mM Na_2HPO_4.

10. Freeze–Thaw Buffer (150 mM NaCl, 50 mM Tris–HCl pH 8.4).
11. Plasmids for AAV production: (A) AAV plasmid containing transgene cassette (promoter and enhancers, gene of interest, poly A tail) flanked by viral inverted terminal repeats (ITRs), in

this study we use AAV encoding Fluc; (B) AAV rep/cap plasmid encoding the replication proteins of AAV and the serotype capsid (in this protocol we use AAV serotype 9); (C) a helper plasmid encoding Ad helper genes necessary for AAV replication (*see* **Notes 1** and **2**).

12. CO_2 incubator for 293T tissue culture and vector production.
13. Tabletop centrifuge.

2.2 Iodixanol Gradient Purification

1. Dry ice and ethanol.
2. Benzonase.
3. Iodixanol for density gradient centrifugation (known as OptiPrep).
4. Type 70Ti rotor ultracentrifuge tubes (39 ml max vol).
5. 16 gauge needles.
6. 10 ml syringes.
7. Beckman Coulter Optima™ L-90K Ultracentrifuge with 70Ti fixed angle rotor.
8. Preparing iodixanol gradient solutions:

0.5 % Phenol Red: 0.25 g in 50 mL 50 % EtOH.

15 % Iodixanol: 12.5 mL 60 % Solution + 5 mL 10× PBS + 10 mL 5 M NaCl + 50 μl

1 M MgCl + 125 μl 1 M KCl + 75 μl 0.5 % phenol red + H_2O up to 50 mL.

25 % Iodixanol: 20.83 mL 60 % iodixanol solution + 5 mL 10× PBS + 50 μl.

1 M $MgCl_2$ + 125 μl 1 M KCl + 100 μl 0.5 % phenol red + H_2O up to 50 mL.

40 % Iodixanol: 33.4 mL 60 % iodixanol solution + 5 mL 10× PBS + 50 μl.

1 M $MgCl_2$ + 125 μl 1 M KCl + H_2O up to 50 mL.

60 % Iodixanol: 25 mL 60 % iodixanol solution + 25 μl.

1 M $MgCl_2$ + 65.5 μl 1 M KCl + 12.5 μl 0.5 % phenol red.

2.3 Virus Concentration

1. Buffer B (20 mM Tris–HCl, 500 mM NaCl, pH 8.5).
2. Amicon Ultra 100 kDa MWCO filters (Millipore, UFC910096).
3. Pierce Slide-A-Lyzer MINI Dialysis Device, 20 kDa MWCO, 2 mL.
4. 0.22 μm Millex-GV Filter Unit filters (Millipore, slgv004sl, hold up volume <10 μl).

2.4 Quantitative PCR (QPCR)

1. Nuclease-free water.
2. AAV primers and Taqman probe which bind to the PolyA tail region in transgene cassette.

Primers sequence:
F2: 5′CCTCGACTGTGCCTTCTAG3′
R2: 5′TGCGATGCAATTTCCTCAT3′
Probe sequence: 5′-6FAM-TGCCAGCCATCTGTTGTTTGCC—MGB (minor groove binding)

3. QPCR master mix.
4. 96 well plate.
5. Plate covers.
6. Applied Biosystems 7500 quantitative PCR thermal cycler.

2.5 Intravenous Injection Components

1. Mouse restrainer.
2. 40 °C water in a beaker.
3. 28 gauge ½ in. insulin syringes.
4. 70 % alcohol wipes.
5. Sterile gauze.

2.6 Luciferase Substrate

D-luciferin sodium or potassium salt. Suspend lyophilized powder to 25 mg/ml in sterile PBS pH 7.2–7.4. Store 0.5–1.5 ml aliquots at −80 °C protected from light.

2.7 Anesthesia and Imaging System

1. Perkin Elmer IVIS® Spectrum optical imaging system.
2. XGI-8 gas anesthesia system.
3. Living Image® software.

3 Methods

3.1 Production, Purification, and Quantitation of AAV Vector

Day 1:

Plate fifteen 15 cm plates with 1×10^7 293T cells per plate in DMEM 10 % FBS, 1 % p/s (20 ml of media/plate).

Day 2: *See* **Note 3**.

1–2 h before transfection, replace media with 20 ml of fresh DMEM 10 % FBS, 1 % Pen/Strep.

Prepare transfection mix (below is for 1 plate).

Tube A	Amount/volume (μg)	Tube B	Volume
AAV plasmid w/gene of interest	9.94	2× HEBS (*see* **Note 4**)	780 μl
Ad helper plasmid	25.56		
rep/cap plasmid	12.37		
2 M CaCl	96.9 μl		
2.5 mM HEPES	Up to 780 μl		
Total volume	780 μl	Total volume	780 μl

Add Tube A to Tube B drop-wise while vortexing Tube B. Vortex for 1 min.

Incubate mix at room temperature for 20 min. A slightly cloudy mixture should be visible.

Add ~1.5 ml of virus mix per 15 cm plate, distributing drop-wise over the surface of the plate. Tilt plate side to side, front to back, to mix evenly and incubate overnight.

Day 3:

Replace media with DMEM 2 % FBS 1 % Pen/Strep.

Day 5:

Aspirate ½ plate media volume (~10 ml) from each plate.

Collect cells from plates by washing them off with remaining media or gently using a cell scraper. Spin at 400 × *g* for 5 min.

Gently remove supernatant. Loosen cells by flicking bottom of tube and resuspend cells in 1 ml EDTA PBS per plate of cells. Spin at 400 × *g* for 5 min.

Gently remove supernatant. Resuspend cells in 1 ml lysis buffer per plate of cells. Store at −80 °C.

Day 6:

Iodixanol Gradient Purification

1. Complete three freeze–thaw cycles with AAV lysate using a 37 °C water bath and a dry ice ethanol bath. Vortex after each thaw.
2. Add Benzonase for a final concentration of 50 U/ml to lysate and incubate at 37 °C for 30 min.
3. Centrifuge at 4,000 × *g* for 20 min at room temperature. Collect vector-containing supernatant and discard pellet.
4. Prepare gradient in 70 Ti tube (you may either use a pump, in which case you layer the gradient underneath the virus fraction going from least to most dense or you can use a syringe and carefully layer the virus fraction on top going from most to least dense) (*see* **Note 5**). The gradient volumes are as follows:

Add virus in clarified lysate (~15 ml) to bottom of Type 70Ti rotor quick seal tube.

9 ml of 15 % iodixanol

6 ml of 25 % iodixanol

5 ml of 40 % iodixanol

5 ml of 60 % iodixanol

Fill to top with lysis buffer

5. Seal top of tube.
6. Spin at 70,000 RPM (360,562 RCF_{avg}) in type 70Ti rotor for 1 h at 20 °C.

7. Remove ~3 ml from the 40 % iodixanol layer. Do not take the bluish band above the 40 % layer—this is cell debris/proteins. This step is crucial for a clean prep. Insert needle at 40–60 interface with bevel up and slowly withdraw 2 ml. Turn needle bevel down and then take 1 ml.
8. Store at +4 °C if continuing purification next day. Store at −80 °C if purification will not occur next day.

Day 7:

Concentrating sample and buffer exchange to PBS

1. Bring up to 20 ml in Buffer B and concentrate down on Amicon Ultra 100 kDa MWCO filters to 1–2 ml.
2. Wash three times with Buffer B (10 ml each) and concentrate down to 1–2 ml each time.
3. Dialyze at room temp with two buffer changes (2 h each) against PBS (or overnight at 4 °C). Use Pierce Slide-A-Lyzer MINI Dialysis Device, 20 kDa MWCO, 2 mL. Bring to room temperature before filtering.
4. Filter through 0.22 μm Filter Unit.
5. Aliquot and store at −80 °C.

Day 8:

QPCR of AAV genome concentrations (titers):

1. Dilute vector 1:100–1:1,000 in nuclease-free water and vortex.
2. Use in QPCR.

qPCR Protocol

- Use a linearized AAV plasmid as your genome copy (g.c.) standard. Make a 10^7 g.c./μl stock and store in aliquots at −20 °C. Create a standard from 10^7 to 10^2 g.c./mL.
- Prepare the following master mix. Make enough to measure the standards and samples in triplicate:
  - 1×
  - 2 μl H_2O
  - 1.2 μl primer mix (F and R = 5 μl of each 100 μm stock in 90 μl of water)
  - 0.8 μl TM FAM Probe 2.5 μM
  - 5 μl of TaqMan Fast universal PCR Master Mix 2×
- Mix and aliquot 9 μl in wells of plate
- Add 1 μl of template
- Setup program on the thermal cycler:

- Stage 1: Reps 1, 95 °C: 20 s
- Stage 2: Reps 40, 95 °C: 03 s, 60 °C: 30 s
- Sample volume 10 μl
- Collect data at stage 2, step 2
- Save program, hit start.

- Analyze data. Slope for standard should be ~–3.3.
  - Multiply g.c. calculated in qPCR report by dilution factor ×1000 (microliters to milliliter) ×2 (for matching single stranded AAV to plasmid standard) to calculate g.c./ml.

3.2 Intravenous Injection of AAV into Mice

1. Load syringe with AAV-Fluc vector (typically in PBS). Remove all air bubbles. Volumes to inject into the tail vein in mice are usually between 50 and 300 μl.
2. Place mouse in restrainer and adjust for a snug fit, with the tail hanging out the opening. Place gentle tension on tail. This prevents the mouse from trying to turn around in the restrainer.
3. Dip the animal's tail into warm water (*see* **Note 6**) for approximately 30 s to dilate the veins (*see* **Note 7**).
4. Sterilize the tail by thoroughly wiping with an alcohol pad.
5. Grasp the tail between your forefinger and thumb and rotate tail to see one of the two lateral tail veins (*see* **Note 8**).
6. With the other hand, hold the loaded syringe with *bevel facing up* and gently slide the needle in trying not to go too deep (*see* **Note 9**).
7. Once you are inside the vein, gently push on the plunger on the syringe (*see* **Note 10**). The blood color will flush out of vein as your clear solution (PBS) is injected.
8. Remove needle and clamp the injection site with sterile gauze for 30 s or so until the bleeding stops.
9. Place mice back in cage.

3.3 Bioluminescence Imaging of AAV Encoded Fluc

1. The timing of imaging depends on the serotype of AAV, the vector genome components, as well as the dose injected. For example expression kinetics in vivo are slow for AAV2 and strong signal may take over 4 weeks. On the other hand AAV8 and AAV9 have rapid expression and we have observed robust bioluminescence 3 days after i.v. injection. Self-complementary AAV vectors have more rapid and robust expression kinetics compared to standard single stranded AAV [13] in vivo. Finally doses for local injections of mice (Fig. 1a) (~20 g) tend to be in the 10^{10}–10^{11} g.c. range while systemic injections (Fig. 1b) are >10^{12} g.c. and researchers tend to use self-complementary vectors.

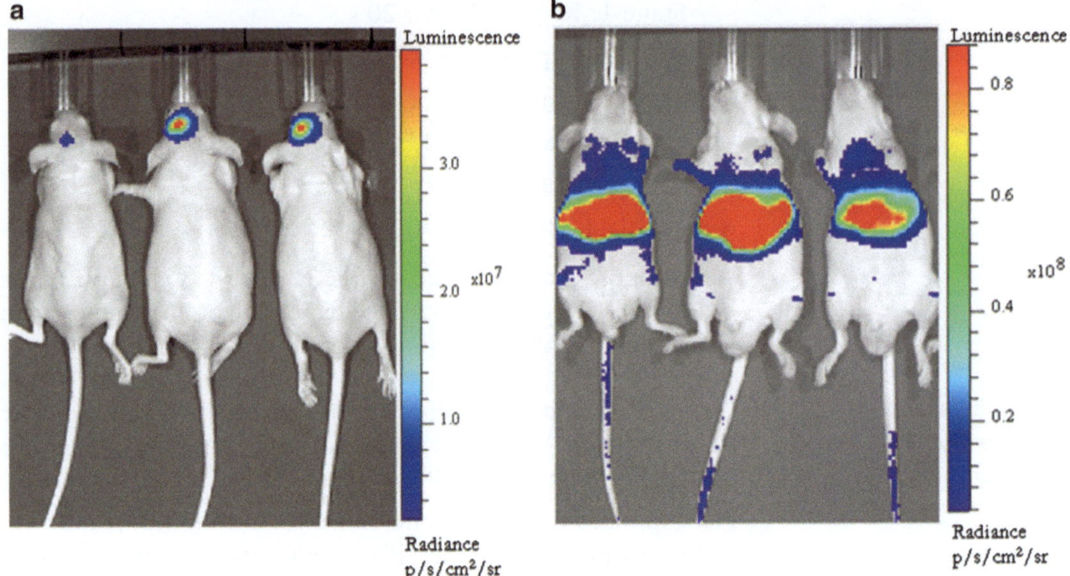

Fig. 1 Typical presentation of AAV-mediated Fluc bioluminescence after direct intracranial injection (**a**) or tail vein injection (**b**) of vector. Note the intense signal localized to the brain in (**a**), while the signal after i.v. injection is most intense from liver (**b**)

2. Make sure the bioluminescent imager is on. The following will describe the procedure used for an IVIS® Spectrum imager with a PC running Living Image® software.
3. Open Living Image software and click "Initialize" to begin cooling the camera. Click the red temperature bar and set stage temperature to 37 °C.
4. For the anesthesia station, first fill the isoflurane tank. Then turn on oxygen tank, the oxygen switch on the station, the pump and open the valves for the induction chamber (the box where mice are anesthetized) as well as the IVIS cabinet. Turn on the isoflurane to 2.5 and adjust flow to 0.5 in the cabinet and 2.5 in the induction chamber.
5. After 5 min, place mice (up to 5) in the induction chamber. Wait until they are completely unconscious.
6. Once the camera reaches the operating temperature (−90 °C) imaging can proceed.
7. Inject the mice intraperitoneally (i.p.) with 150 mg/kg of D-luciferin (for a 25 g mouse, this is 3.75 mg, or 150 μl of the stock solution). Inject all mice as rapidly as possible so that the time point 0 is close for all. Don't inject more than five mice at a time which is the maximum that the stage holds (*see* **Note 11**).

8. Place mice into the imaging cabinet and carefully position their nose into anesthesia cones.
9. After the desired substrate incubation time (5–10 min) begin data acquisition.

Camera settings:

Luminescent and Photograph boxes should be checked.

Set the acquisition to auto-acquire. This will ensure that no saturating signals occur (*see* **Note 12**).

Luminescence settings:

- Binning is typically set to medium (the higher the binning the lower the spatial resolution but higher signal and vice versa)
- F/Stop = 1 (lower F/stop = larger aperture and more light but lower resolution; higher F/stop = 8 less light but sharper image if signal is high enough)
- Emission filter = open
- Subject height = 1.5 cm
- Field of view: C = 1 mouse view; D = 5 mice view

Photograph settings:

- Binning set to medium
- F stop = 8

10. Image each group of mice dorsal (belly down) and ventral (belly up). This gives the best information about transduction in different organs. For instance, image ventrally for liver transduction (Fig. 2a) and dorsally for brain transduction (Fig. 2b) (*see* **Note 13**).

3.4 Data Analysis

1. Go to File > Browse
2. Select file
3. To load multiple images for comparison, select "add to list"
4. Go to File> Browse and select file (continue to add the rest of files).
5. Highlight all selected files in browser window.
6. Click "load as group"
7. Go to image palette and select image adjust tab
8. One can normalize all images to same scale by selecting "auto" on color limit scale

One can also manually adjust min/max of image (*see* **Note 14**).

To quantitate signal in a particular region of interest (ROI), open an image and select the ROI tools tab. Select drawing tools of interest (e.g., oval) and draw around each animal with the same size and location. Select measure ROIs to get luminescence values and then export it to excel file (*see* **Notes 15–17**).

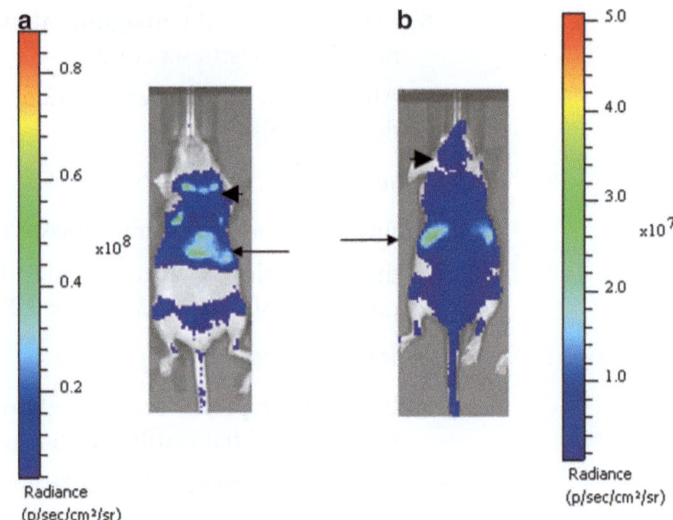

Fig. 2 Imaging dorsally and ventrally to obtain more transduction information. The same mouse injected i.v. with AAV-Fluc was imaged ventrally (**a**) and dorsally (**b**). Note in (**a**), liver (*arrow*) as well as lung/heart (*arrowhead*) localized signal is observed. Note in (**b**), liver signal can also be observed (*arrow*) as well as signal from the head (*arrowhead*)

4 Notes

1. AAV is very stable at −80 °C although it can aggregate after freezing–thawing. For best results store AAV in small aliquots and use a new vial for each experiment.
2. AAV titers are usually expressed as genome copies/ml (g.c./ml). This is a physical titration method and does not provide information on functional activity of transgene expression. However, in general, preparations have similar functional activity from a given vector provider, so comparison of different vectors can be dose matched using the g.c./ml value. Typical titers of AAV preparations for small animal work range from 10^{11} to 10^{13} g.c./ml and usually are in the mid to high 10^{12} g.c./ml range.
3. High efficiency transfection (at least 80 %) is a key parameter for high yield AAV production. 293T cells transfect best at around 70 % confluency with this protocol.
4. The pH of the 2× HEBS is crucial (should be 7.0–7.05) for high levels of transfection efficiency. Triple-check the pH of this solution and make sure your pH meter is accurate and calibrated!
5. It is best to load a maximum of 7.5 plates per gradient (i.e., 2 gradients per 15 plate prep) when using the 39 ml tubes as this is around the max resolving (best for purity) capacity of the gradient for that much cell lysate.

6. The temperature of the water is very important. Use a thermometer to accurately monitor it. Heat the water in a microwave. Keep the temperature between 38 and 42 °C. Pulse in microwave as needed to maintain temperature between injections. Be careful! Heating the mouse tail over 43 °C may severely burn it and injure the animal.
7. The warm water helps to dilate the tail vein making it much easier to see. It is much easier to see on mice with pink skin (e.g., nude mice) compared to mice with dark skin (e.g., C57BL/6).
8. If it is difficult to inject the solution into the tail vein try the other vein on the other side and start fresh.
9. Start the injection as close to the tip of the tail as possible. This serves two purposes. First, even though the vein is smaller at the tip it is easier to locate than at the base. Second, if you miss the vein you can work up towards the base without having the sample leak out a higher prior injection site.
10. If there is a lot of resistance, stop pushing. You are most likely in the tissue around the vein. If this happens, retract needle and go to a new insertion site.
11. Maximum signal for Fluc generally occurs 5–10 min post i.p. injection of substrate. The signal is fairly stable and decays slowly over approximately 1 h. However for consistency, image all groups of mice exactly the same time-post substrate injection.
12. If one or more mice have extremely high signal compared to the others take one acquisition including all of the mice. Then remove the mice with high signal and do another data acquisition. This will allow one to visibly detect the signal on the mice with lower signal, while retaining a non-saturated reading for the mice with high signal.
13. Although sensitive with low background, BLI has its limitations. For example, tissue and hemoglobin absorb much of the green light emitted by Fluc. More red-shifted reporters can minimize this. Imaging through the skull or backbone is also difficult. The 1 mm thick skull reduces signal going to the camera by one log.
14. The software automatically adjusts the signal scale based on the most intense signal. Lower signals may not be visible on automatically adjusted scale. Adjusting the signal threshold may allow one to observe lower transduced areas, including the brain (*see* Fig. 3).
15. Bioluminescent signal can be expressed as photons/second (i.e., total flux), or in photons/s/cm^2/sr (i.e., Radiance). Radiance takes into account the area (cm^2) of light emission from the surface of the subject as well as the "cone-like" emission shape (sr, steradian) emitted from the subject surface.

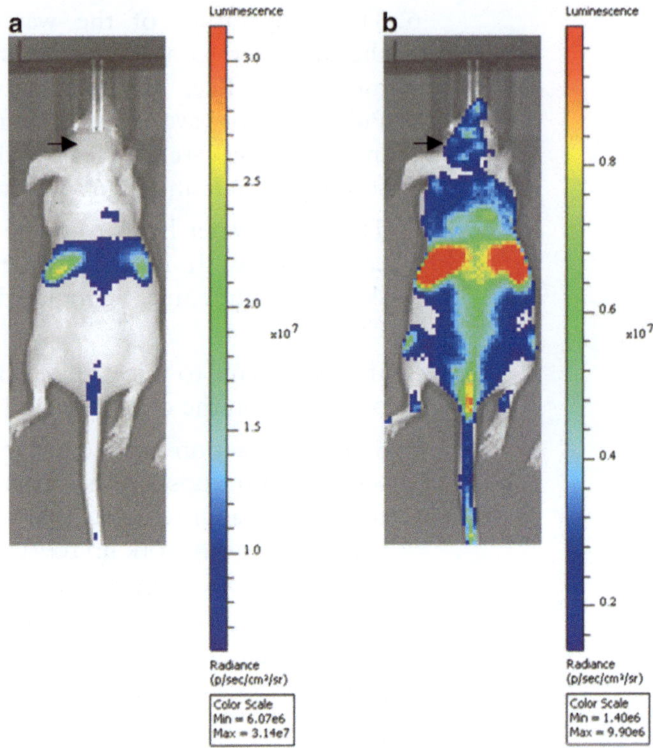

Fig. 3 Adjusting signal scale in software to visualize areas with lower signal. (**a**) Automatically adjusted image. (**b**) Scale adjusted to visualize transduction in head region

16. To confirm measured luminescence is above background, it is best measure luminescence of naïve mice. Acquire a BLI image and photograph of the naïve mouse and draw an ROI. Typical background levels on nude mice have a radiance of less than 10^4 photons/s/cm^2/sr. Levels much below should be cautiously interpreted as bona fide signal.
17. Although BLI can be useful for differentiating transduction of large organs from one another (e.g., liver vs. heart, brain vs. lung) the spatial resolution to differentiate different regions and depths in the brain is limited. For example if there is transduction of tissue near the surface of the brain in one group of mice and an equal level of transduction in a deeper region in another group of mice the signal will be more intense in the first group. Therefore to find the precise region of transduction as well as the transduced cell type it is recommended to confirm transduction post euthanasia using luciferase assays of homogenized tissues as well as histological analysis of tissue sections. For instance if one is interested in brain transduction of AAV, sacrifice the animals imaged by BLI, dissect specific regions of the brain (e.g., striatum, endothelial cells, cerebellum)

and homogenize the tissue. Perform a luciferase assay in a luminometer to determine luciferase activity in tissue and normalize to protein amount with a Bradford assay to get relative light units/mg of tissue (RLU/mg). To determine transduced cell type it is best to do the experiment with an AAV vector encoding GFP, and perform immunostaining for cell specific markers.

Acknowledgement

This work was supported by NIH/NINDS 1 R21 NS081374-01. CM has a financial interest in Exosome Diagnostics, Inc. CM's interests were reviewed and are managed by the Massachusetts General Hospital and Partners HealthCare in accordance with their conflict of interest policies.

References

1. Chung E, Yamashita H, Au P et al (2009) Secreted Gaussia luciferase as a biomarker for monitoring tumor progression and treatment response of systemic metastases. PLoS One 4(12):e8316
2. Peterson JR, Infanger DW, Braga VA et al (2008) Longitudinal noninvasive monitoring of transcription factor activation in cardiovascular regulatory nuclei using bioluminescence imaging. Physiol Genomics 33(2):292–299
3. Takakura H, Hattori M, Takeuchi M, Ozawa T (2012) Visualization and quantitative analysis of G protein-coupled receptor-beta-arrestin interaction in single cells and specific organs of living mice using split luciferase complementation. ACS Chem Biol 7(5):901–910
4. Zincarelli C, Soltys S, Rengo G, Rabinowitz JE (2008) Analysis of AAV serotypes 1-9 mediated gene expression and tropism in mice after systemic injection. Mol Ther 16(6):1073–1080
5. Ignowski JM, Schaffer DV (2004) Kinetic analysis and modeling of firefly luciferase as a quantitative reporter gene in live mammalian cells. Biotechnol Bioeng 86(7):827–834
6. Corish P, Tyler-Smith C (1999) Attenuation of green fluorescent protein half-life in mammalian cells. Protein Eng 12(12):1035–1040
7. Coura Rdos S, Nardi NB (2007) The state of the art of adeno-associated virus-based vectors in gene therapy. Virol J 4:99
8. Samaranch L, Salegio EA, San Sebastian W et al (2013) Adeno-associated virus serotype 9 transduction in the central nervous system of nonhuman primates. Hum Gene Ther 23(4): 382–389
9. Burger C, Gorbatyuk OS, Velardo MJ et al (2004) Recombinant AAV viral vectors pseudotyped with viral capsids from serotypes 1, 2, and 5 display differential efficiency and cell tropism after delivery to different regions of the central nervous system. Mol Ther 10(2): 302–317
10. Foust KD, Nurre E, Montgomery CL et al (2009) Intravascular AAV9 preferentially targets neonatal neurons and adult astrocytes. Nat Biotechnol 27(1):59–65
11. Zhang H, Yang B, Mu X et al (2011) Several rAAV vectors efficiently cross the blood-brain barrier and transduce neurons and astrocytes in the neonatal mouse central nervous system. Mol Ther 19(8):1440–1448
12. Calcedo R, Vandenberghe LH, Gao G et al (2009) Worldwide epidemiology of neutralizing antibodies to adeno-associated viruses. J Infect Dis 199(3):381–390
13. McCarty DM, Fu H, Monahan PE et al (2003) Adeno-associated virus terminal repeat (TR) mutant generates self-complementary vectors to overcome the rate-limiting step to transduction in vivo. Gene Ther 10(26):2112–2118

Chapter 17

Simultaneous In Vivo Monitoring of Regulatory and Effector T Lymphocytes Using Secreted Gaussia Luciferase, Firefly Luciferase, and Secreted Alkaline Phosphatase

Grant K. Lewandrowski, Ciara N. Magee, Marwan Mounayar, Bakhos A. Tannous, and Jamil Azzi

Abstract

Regulatory T cells (T_{regs}) are amongst the most widely studied cells in a variety of immune-mediated conditions, including transplantation and Graft Versus Host Disease (GVHD), cancer and autoimmunity; indeed, there is great interest in the tolerogenic potential of T_{reg}-based therapy. Consequently, the need to establish the mechanisms that determine T_{reg} survival and longevity, in addition to developing new tools to monitor these parameters, is paramount. Using both a mouse model of GVHD and a mouse model of Type 1 Diabetes (T1D), we describe herein a dual reporter system based on Gluc and multiplexed with SEAP and non-secreted Firefly luciferase (Fluc), which permits simultaneous imaging and noninvasive tracking of two different T-cell populations ($CD4^+CD25^+$ T_{regs} and $CD4^+CD25^-$ T_{con} cells) in vivo by transducing the cells with different lentiviruses bearing distinct color signatures. This new technology promises to overcome the limitations of the conventional methods currently available to study lymphocyte survival in vivo. Furthermore, this novel technique has applications not only in autoimmunity and alloimmunity, but also in the wider field of immunology.

Key words Luciferase, Bioluminescence imaging, Lymphocytes, Tregs, Homeostasis, Survival, GVHD, Type 1 diabetes

1 Introduction

The discovery of $CD4^+CD25^+$ regulatory T cells (T_{regs}) in 1995 was accompanied by compelling evidence of their critical role in the maintenance of self-tolerance and prevention of autoimmunity [1]. These cells, characterized by the expression of the transcription factor FoxP3, comprise approximately 5–10 % of $CD4^+$ T cells

Grant K. Lewandrowski and Ciara N. Magee contributed equally to this work.

Christian E. Badr (ed.), *Bioluminescent Imaging: Methods and Protocols*, Methods in Molecular Biology, vol. 1098,
DOI 10.1007/978-1-62703-718-1_17, © Springer Science+Business Media New York 2014

in normal naive mice and are of thymic origin (natural T_{regs}; nT_{regs}) [1, 2]. Studies have shown that it is possible to induce T_{regs} from naïve peripheral $CD4^{+}CD25^{-}$ T cells, depending on the stimulus and cytokine milieu [3], and that these adaptive T_{regs} (iT_{regs}) also possess the ability to regulate and suppress the immune response, even in the absence of nT_{regs} [4]. T_{reg} therapy has emerged as a promising tolerogenic approach for a variety of immune-mediated conditions, including autoimmune disease and Graft-Versus-Host-Disease (GVHD); in transplantation models, T_{regs} have been shown to actively regulate allogeneic T effector (T_{eff}) cell responses and promote transplantation tolerance [5, 6]. Although our knowledge of the wide spectrum of their suppressive properties is constantly expanding [7, 8], the homeostatic mechanisms involved in maintaining the population and functions of T_{regs} have been largely unexplored [9]. Given the increasing number of clinical trials testing the safety and efficacy of T_{reg}-based immunotherapy [10], the need to establish the mechanisms that determine the survival and longevity of T_{regs}, and, consequently, the means of monitoring these parameters, is paramount.

Secreted reporters in the blood or serum have been shown to be important tools for the detection, quantitation and noninvasive monitoring of different biological processes, including cell tracking, in experimental models [11–18]; indeed, they have several advantages over conventional in vivo imaging methods for monitoring gene expression [19, 20]. In culture, secreted reporters can be quantified in cell-free conditioned medium without the need for cell lysis, thereby leaving the cells intact and available for confirmation analysis. In vivo, secreted reporters can be easily monitored in animals by withdrawing a few microliters of blood or urine and assaying for their activity, allowing real-time monitoring of in vivo processes; furthermore, bioluminescence imaging can be performed to localize cell activity in vivo, although this method requires expensive instrumentation (cooled CCD camera), is time-consuming, involves frequent anesthesia and repeated systemic substrate injections, and is limited by photon absorption by tissues as a function of depth [20].

The most commonly used serum reporter is the secreted alkaline phosphatase or SEAP [12, 13, 15–17, 21–25]. Normally, alkaline phosphatase is not secreted; however, it has been shown that placental alkaline phosphatase is expressed and released efficiently into the conditioned medium of mammalian cells that have been engineered to express this reporter [21]. The level of SEAP in the conditioned medium is linearly related to the changes in its intracellular mRNA level as well as cell number [21, 22]. It is also possible to quantitatively detect SEAP in the serum of animals. In 1997, Abruzzese et al. used the SEAP serum level to monitor the induction of gene expression followed by gene transfer to the muscles of adult mice [26]. Since then, the SEAP reporter assay has been used as a surrogate serum marker to monitor different in vivo processes including tumor growth and drug efficacy [13], DNA vaccination [27];

promoter and transcription factors activation and inhibition [12, 16, 25]; viral gene transfer [28]; inflammatory events [17]; viral infection and replication [29]; and endoplasmic reticulum stress [23].

We have recently established and reported a novel secreted luciferase reporter from the marine copepod *Gaussia princeps* known as Gaussia luciferase (Gluc) [11, 14, 18, 30, 31]. We have shown that this reporter is naturally secreted from mammalian cells in an active form and that the levels of Gluc in the conditioned medium correlate linearly with respect to cell number, growth and proliferation [11, 31]. The Gluc reporter system has several advantages over the SEAP assay including a much reduced assay time (1–10 min vs. 1.5–2 h), greatly increased sensitivity (20,000-fold in vitro and 1,000-fold in vivo), and a linear range covering over 5 orders of magnitude of cell number. In vivo, the Gluc reporter assay has been used to localize tumors and to monitor tumor formation and proliferation using bioluminescence imaging [31, 32]. Since Gluc is naturally secreted, we have shown that its level in the blood or urine can be used as a marker to monitor different in vivo biological events such as tumor growth and response to therapy, viral infection and replication as well as the viability of circulating cells [18]. This is a crucial advantage when investigating the behavior of widely circulating cells where a lack of focal accumulation/cell density precludes the use of typical imaging techniques to localize the cells; in these instances, blood or serum levels can be used to reflect cell activity.

We describe herein a dual reporter system based on Gluc and multiplexed with SEAP and non-secreted Firefly luciferase (Fluc) which permits simultaneous imaging and noninvasive tracking of two different T-cell populations ($CD4^{+}CD25^{+}$ T_{regs} and $CD4^{+}CD25^{-}$ T_{con} cells) in vivo by transducing the cells with different lentiviruses bearing distinct color signatures. One further advantage offered by this method of cell identification is that even if the cell loses its normal identifier due to cell plasticity, it is still possible to identify the original lineage of the cell, as it will retain its transduced color signature. This has become particularly important as we become more aware of cell plasticity, e.g., induced T_{regs} arising from conventional T cell precursors can lose FoxP3 expression and regulatory function over time, becoming effector T cells [33].

2 Materials

2.1 Large-Scale Lentiviral Packaging and Concentration

1. Dulbecco's Modified Eagle Medium (DMEM): High glucose, HEPES, 10 % Fetal Bovine Serum, 1 % Penicillin/Streptomycin. Store at 4 °C.
2. Opti-MEM® Reduced Serum Medium: GlutaMAX, 1 % Penicillin/Streptomycin. Store at 4 °C.
3. 0.05 % Trypsin–EDTA. Store at 4 °C.
4. Sterile Phosphate Buffered Saline (PBS).

5. 2 M Calcium Chloride solution.
6. 2× HEPES Buffered Saline solution (HeBS): 0.28 M NaCl, 0.05 M HEPES, 1.5 mM Na_2HPO_4 anhydrous, distilled H_2O up to 1 L, pH 7.00 (*see* **Note 1**).
7. 2.5 mM HEPES solution.
8. pCMV-ΔR8.91 plasmid (Addgene, Cambridge, MA, USA).
9. pCMV-VSV-G plasmid (Addgene, Cambridge, MA, USA).
10. pCSCW-Gluc-IRES-GFP plasmid vector (*see* **Note 2**).
11. pCSCW-Fluc-IRES-mCherry plasmid vector (*see* **Note 3**).
12. pCSCW-SEAP-IRES-mCherry plasmid vector (*see* **Note 4**).
13. Standard treated 15 cm cell culture dishes.
14. SW32 Rotor Tubes: Thinwall, Polyallomer, 38.5 mL, 25 × 89 mm.
15. Standard conical tubes (50 mL).
16. PVDF Vacuum Filtration Assembly: High volume, 0.45 μm.
17. Cultured HEK/293T (containing SV40 Large T-antigen) cell line.
18. Standard vortex mixer.
19. Coulter counter or hemacytometer.
20. Trypan blue.
21. Tabletop centrifuge.
22. Microcentrifuge.
23. Ultracentrifuge.
24. SW32 swinging bucket rotor.
25. Confocal Laser Scanning Microscope.

2.2 T Cell Isolation and Activation

1. MACS Buffer: 0.5 % Bovine Serum Albumin, 2 mM EDTA, Sterile PBS. Store at 4 °C.
2. RPMI-1640 Medium: 10 % Fetal Bovine Serum, 1 % Penicillin/ Streptomycin, 1 % L-Glutamine. Store at 4 °C.
3. 70 μm filter units.
4. 500 mL Rapid-Flow Filter Unit (Thermo Fisher, Waltham, MA, USA).
5. Standard conical tubes (15 mL).
6. Syringes (1 mL).
7. ACK (Ammonium-Chloride-Potassium) lysis buffer (GIBCO, Grand Island, NY, USA).
8. MACS LD & MS columns (Miltenyi Biotec, Cambridge, MA, USA).
9. Mouse Regulatory T cell Isolation Kit (Miltenyi Biotec, Cambridge, MA, USA).
10. Dynabeads: Mouse T-Activator CD3/CD28 beads (GIBCO, Grand Island, NY, USA).

2.3 Lentiviral Infection of T Cells

1. Polybrene. Resuspend to 8 mg/mL stock. Store at 4 °C.

2.4 Bone Marrow Extraction

1. Sterile Petri dishes.
2. Dissection scissors.
3. Forceps.
4. Razor blades.
5. Syringes (5 mL).
6. Needles (27 Gauge).
7. CD3ε MicroBead Kit (Miltenyi Biotec, Cambridge, MA, USA).
8. MACS LD columns (Miltenyi Biotec, Cambridge, MA, USA).

2.5 Injection of Cells

1. Magnetic bead separation rack (New England Biolabs, Beverly, MA, USA).
2. U-100 insulin syringes (28.5 Gauge).

2.6 Bioluminescent and Chemiluminescent Assays and Bioluminescent Mouse Imaging

1. Coelenterazine free base 97 % (Nanolight, Pinetop, AZ, USA). Store at −20 °C (*see* **Note 5**).
2. Hydrochloric acid.
3. Methanol.
4. Opaque white 96-well assay plates.
5. Great EscAPe SEAP Reporter System 3 (Clontech, Mountain View, CA, USA).
6. D-Luciferin Potassium Salt (Gold Biotechnology, St Louis, MO, USA). Store at −80 °C (*see* **Note 6**).
7. Nude athymic mice.
8. Isoflurane mouse anesthetic.
9. 50 mM Ethylenediaminetetraacetic acid (EDTA).
10. Microcentrifuge tubes.
11. Luminometer.
12. Cooled charge-coupled device (CCD) camera.
13. XGI-8 gas anesthesia system.
14. Living Image software.

3 Methods

3.1 Large-Scale Lentiviral Packaging and Concentration

3.1.1 Day 1

1. Plate out 2.1×10^7 293T cells in 20 mL of supplemented DMEM medium (*see* **Note 7**) per plate in ten 15 cm plates for each lentiviral vector to be packaged (*see* **Note 8**).

3.1.2 Day 2

1. Aspirate out existing medium from all plates and replace with 20 mL of supplemented DMEM medium one hour prior to transfection.

2. Prepare Calcium-Phosphate transfection mix in two separate 50 mL conical tubes. To Vial 1 add the following: 180 μg desired plasmid vector, 180 μg pCMV-ΔR8.91 plasmid, 120 μg pCMV-VSV-G plasmid, 969 μL 2M $CaCl_2$ solution, and bring total volume up to 7,800 μL with 2.5 mM HEPES solution. To Vial 2 add 7,800 μL 2× HeBS solution.
3. Mix contents of Vial 1 drop-wise to Vial 2 while vortexing at medium speed. Tightly cap Vial 2 without ceasing vortexing and increase vortex to max speed for 1 min (*see* **Note 9**).
4. Incubate Vial 2 at room temperature for 20 min.
5. After incubation, add 1.5 mL of transfection mixture drop-wise to each of ten plates of 293T cells and gently rock plates to evenly distribute the mixture.

3.1.3 Day 3

1. Wash plates by aspirating off existing medium and replacing with 13 mL supplemented Opti-MEM medium (*see* **Note 10**) for each plate. Carefully pipette down sides of plate to avoid disrupting cell monolayer.
2. Aspirate off the 13 mL of supplemented Opti-MEM and add an additional 13 mL of supplemented Opti-MEM to each plate.
3. Return plates to incubator for 24 h.

3.1.4 Day 4

1. Harvest the 13 mL of virus-saturated medium from each plate and store at 4 °C in 50 mL conical tubes. Refresh plates with additional 13 mL supplemented Opti-MEM and incubate for 24 h.

3.1.5 Day 5

1. Harvest 13 mL virus-saturated medium from each plate, combine with harvested medium from Day 4, and centrifuge at 450 × *g* for 5 min.
2. Decant supernatant and filter using a 0.45 μm vacuum filtration assembly. Transfer filtrate to SW32 ultracentrifuge tubes.
3. Ultracentrifuge the filtrate at 133,907 × *g* (max) for an SW32Ti Rotor for 98 min at 4 °C.
4. Aspirate supernatant and resuspend all viral pellets using a total of 200 μL supplemented Opti-MEM. Aliquot and store at −80 °C.

3.2 T Cell Preparation

3.2.1 Day 1: T Cell Isolation and Activation

1. Prepare sterile MACS buffer (*see* **Note 11**).
2. Prepare sterile RPMI-1640 complete culture medium (*see* **Note 12**).
3. Harvest the spleens from the relevant mice using aseptic technique; store them temporarily in a 15 mL conical tube containing sterile PBS.
4. Prepare a single-cell suspension by first placing the cell strainer in a 50 mL conical tube and adding the spleens into the strainer.

5. Mechanically dissociate the spleens through the cell strainer using the flat edge of the syringe plunger. Rinse the cell strainer frequently with sterile PBS.
6. Add sterile PBS to a total volume of 45 mL and centrifuge at 300 × *g* for 6 min.
7. Discard the supernatant and add 1–2 mL of ACK lysis buffer per spleen, making sure to pipette thoroughly until the cell pellet is completely resuspended.
8. Allow solution to rest for 3 min and then terminate lysis reaction by adding sterile PBS to a total volume of 45 mL.
9. Filter this solution through a fresh cell strainer into another 50 mL conical tube and then centrifuge at 300 × *g* for 8 min.
10. Discard the supernatant and resuspend the cell pellet in 10 mL sterile PBS for counting.
11. Centrifuge cells for 8 min at 300 × *g*. Aspirate supernatant completely.
12. Prepare the cells for MACS isolation using the Mouse Regulatory T cell Isolation Kit (Miltenyi Biotec): $CD4^+$ T cells are pre-enriched by depletion of unwanted cells prior to positive selection of $CD25^+$ cells from the $CD4^+$ T cell-enriched fraction.

    Resuspend the cells in 40 μL of MACS buffer per 10^7 cells; add 10 μL of Biotin-Antibody cocktail per 10^7 cells; mix well and refrigerate (4 °C) for 10 min.
13. Add 30 μL of buffer, 20 μL of Anti-Biotin MicroBeads and 10 μL of CD25-PE antibody per 10^7 cells and refrigerate for a further 15 min in the dark.
14. Wash cells by adding 1–2 mL of buffer per 10^7 total cells and centrifuge at 300 × *g* for 8 min. Aspirate supernatant completely.
15. Resuspend cell pellet in 500 μL MACS buffer for up to 1.25 × 108 cells (using an LD column).
16. Place LD Column in the magnetic field of a suitable MACS Separator (for details *see* LD Column data sheet).
17. Prepare column by rinsing with 2 mL of buffer and apply cell suspension onto the column.
18. Collect unlabeled cells which pass through and wash column twice with 1 mL of buffer. Perform washing steps by adding buffer successively once the column reservoir is empty. Collect total effluent. This is the unlabeled CD4+ T cell fraction.
19. Centrifuge isolated $CD4^+$ T cells at 300 × *g* for 8 min. Aspirate supernatant completely.
20. Resuspend cell pellet in 90 μL of buffer and add 10 μL of Anti-PE MicroBeads.

21. Mix well and refrigerate for 15 min in the dark (4 °C).
22. Wash cells by adding 1–2 mL of buffer and centrifuge at 300 × *g* for 8 min. Aspirate supernatant completely.
23. Resuspend up to 10^8 cells in 500 μL of buffer.
24. Place MS Column in the magnetic field of a suitable MACS Separator (for details *see* MS Column data sheet).
25. Prepare column by rinsing with 500 μL of buffer and apply cell suspension onto the column.
26. Let cells pass through and wash column three times with 500 μL of buffer. These cells are the T_{conv} $CD4^{+}CD25^{-}$ fraction.
27. Remove column from the separator and place it on a suitable collection tube.
28. Pipette 1 mL of buffer onto the column. Immediately flush out the magnetically labeled cells (CD4+CD25+ cells) by firmly pushing the plunger into the column.
29. Centrifuge the cells at 300 × *g* for 8 min and resuspend in sterile PBS for counting.
30. Wash the Dynabeads with 2 mL RPMI-1640.
31. Add 5×10^4 purified cells in 50 μl RPMI-1640 and 1.3 μl of washed Dynabeads (a 1:1 ratio of cells to Dynabeads) per well in a 96-well flat-bottomed plate.
32. Incubate cells with Dynabeads at 37 °C overnight.

3.2.2 Day 2: Total Body Irradiation of Host Mice

1. Irradiate the host mice using a rotating carousel, administering a total dose of 9 Gy (900 rads).
2. Allow the mice to recover for at least 3–4 h following irradiation.

3.2.3 Day 2: Lentiviral Infection of Cells of Interest

1. Prepare 1 mL of a 1:1,000 dilution of polybrene 8 mg/mL stock using previously prepared supplemented RPMI-1640 medium to create an 8 μg/mL polybrene solution.
2. Prepare two separate master infection mixes, the first using only the pCSCW-Gluc-IGFP virus and the second using a mixture of the pCSCW-Fluc-ImCherry and pCSCW-SEAP-ImCherry viruses (*see* **Note 13**). Vortex master mixes thoroughly.
3. Pipette 10 μL of master mix per well to the respective wells, in addition to the medium currently in each well (approximately 50 μL). T_{con} cells will receive the pCSCW-Fluc-ImCherry and pCSCW-SEAP-ImCherry virus mix. T_{regs} will receive only the pCSCW-Gluc-IGFP virus.
4. Centrifuge plates in a table-top centrifuge for 2 h at 700 × *g* at 32 °C.
5. Return plates to an incubator for 6 h to recover, at which point the cells are ready for mouse model injections.

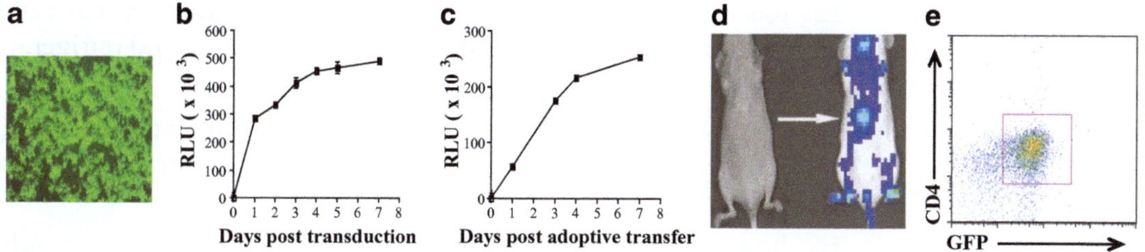

Fig. 1 Isolated NOD CD4+ CD25− T_{con} cells were transduced with lentivirus vector encoding Gluc and GFP, were stimulated in vitro with anti-CD3/CD28 and adoptively transferred into NOD-SCID mouse. (**a**) Fluorescent microscopy image (×20) of transduced CD4+ T cells showing GFP expression. (**b**) Gluc activity in the medium of stimulated CD4+ T cells in vitro. (**c**) Gluc activity measured in the blood of the injected NOD-SCID mouse. (**d**) Bioluminescent imaging of the NOD-SCID mouse that received the transduced CD4+ T cells (*right*) showing a clear signal in the pancreatic area (*arrow*) compared to the negative control mouse (*left*). (**e**) FACS analysis of cells isolated from the pancreatic LNs, confirming the presence of a significant population of T_{regs}

6. Expected results for 24 h post-infection analysis of cells not used in injections demonstrate >90 % transduction efficiency of $CD4^{+}CD25^{-}$ T cells when examined under a fluorescent microscope for CSCW-Gluc-IGFP expression, for example (Fig. 1a).

3.2.4 Day 2: Preparation of Bone Marrow

1. While the irradiated mice are recovering, prepare the bone marrow cells that will support the injected T cells in vivo.
2. Using aseptic technique, dissect legs, removing whole leg with minimal skin. Place on a sterile petri dish and remove muscles and connective tissue. Disarticulate at the tibio-femoral joint.
3. Sterilize a razor blade with EtOH, followed by sterile PBS.
4. On a clean petri dish, cut the femur with a razor blade close to the femoral ball and the lower articulation to open the bone marrow canal. Place the prepared bone on a new petri dish containing sterile PBS.
5. Repeat the above steps to prepare the tibia.
6. Using a 5 mL syringe and a 27G needle, insert the tip of the needle into the bone marrow canal and flush with 1.5 mL of supplemented RPMI-1640 medium (*see* **Note 14**).
7. Pass the resulting suspension through a cell filter into a 50 mL conical tube. Add medium to a total volume of 45 mL.
8. Centrifuge cells for 8 min at 400 × *g*.
9. Discard supernatant and resuspend pellet in 10 mL of sterile PBS for counting.
10. Centrifuge cells for 8 min at 300 × *g*. Aspirate supernatant completely.
11. Prepare the cells for MACS isolation using the Mouse CD3ε Microbead Kit (Miltenyi Biotec): the cell suspension is depleted of CD3ε+ T cells to limit the extent of GVHD induced in the host mice.

12. Resuspend the cells in 100 μL of MACS buffer per 10^7 cells; add 10 μL of CD3ε-Biotin per 10^7 cells; mix well and refrigerate (4 °C) for 10 min.
13. Wash cells by adding 1–2 mL of MACS buffer per 10^7 total cells and centrifuge at $300 \times g$ for 8 min. Aspirate supernatant completely.
14. Resuspend cell pellet in 80 μL MACS buffer per 10^7 cells. Add 20 μL Anti-Biotin MicroBeads per 10^7 total cells.
15. Mix well and refrigerate (4 °C) for 15 min.
16. Wash cells by adding 1–2 mL of buffer and centrifuge at $300 \times g$ for 8 min. Aspirate supernatant completely.
17. Resuspend up to 1.25×10^8 cells in 500 μL of buffer.
18. Place LD Column in the magnetic field of a suitable MACS Separator (for details *see* LD Column data sheet).
19. Prepare column by rinsing with 2 mL of buffer and apply cell suspension onto the column.
20. Perform washing steps by adding buffer successively once the column reservoir is empty. Collect total effluent—this is the unlabeled (CD3ε$^-$) cell fraction.
21. Centrifuge the effluent at $300 \times g$ for 8 min and resuspend in sterile PBS for counting.

3.2.5 Day 2: Injection of Cells

1. Using a pipette, manually pipette up and down into each individual well of the 96-well plates (from step in Subheading 3.2.3) to prevent cells from adhering to the plate bottom.
2. Pool the infected cell-containing medium from these wells (5×10^4 cells per well) into microcentrifuge tubes, ensuring to prevent cross-contamination between different cells infected by separate viruses.
3. Using a magnetic separation rack, pipette out the infected cell-containing medium from the metallic activation beads and transfer to new microcentrifuge tubes, ensuring to prevent cross-contamination between different cells infected by separate viruses.
4. Centrifuge the infected cell-containing medium at $180 \times g$ and discard the supernatant medium, yielding a remaining cell pellet.
5. Resuspend cell pellets in 200 μL sterile 1× PBS, ensuring to prevent cross-contamination between different cells infected by separate viruses.
6. Inject 2×10^6 cells intravenously per mouse into mouse models via the retro-orbital administration route (*see* **Note 15**).

3.2.6 Days 3, 5, 7, 9, 14: Whole Blood Collection and Serum Separation

1. On each of days 3, 5, 7, 9, and 14, for each mouse, perform a minor tail vein cut to collect 40 μL of whole blood, combine with 5 μL of 50 mM EDTA in a microcentrifuge tube and mix thoroughly to prevent clotting (*see* **Note 16**).
2. Store 10 μL of each collected blood sample at 4 °C for Gluc blood secretion assaying on days 7 and 14. Gluc is stable and will not break down in a 1 week time period.
3. In a microcentrifuge, spin the remaining 35 μL of each collected blood sample at 200 × *g* for 5 min. Carefully pipette out and store the supernatant serum at −80 °C for SEAP blood secretion assaying on day 14. SEAP is stable in serum when kept at this temperature.

3.2.7 Days 7 and 14: Bioluminescent Assay Analysis of Gluc Secretion

1. Prepare a solution using 5 mg/mL stock coelenterazine and diluting 1:100 in 1× PBS; 100 μL of this solution will be required for each sample being assayed. Wrap solution container in aluminum foil to prevent light exposure.
2. Stabilize coelenterazine solution by incubating for 30 min at room temperature.
3. Transfer 5 μL of collected whole blood samples per well to a 96-well microtiter plate (*see* **Note 17**). As 10 μL of whole blood was stored per mouse per day, duplicates may be assayed.
4. Using a plate luminometer, set to inject 100 μL of prepared coelenterazine solution per well, measure Gluc bioluminescent activity by acquiring photon counts for 10 s.
5. Expected results will demonstrate the production of Gluc by live proliferating cells to be increased over time (Fig. 1c), showing that this method could be used to monitor T cell viability and proliferation in vivo in real-time. Additionally, at different time points, aliquots of medium taken directly from infected cells could be analyzed for Gluc activity to demonstrate increased Gluc production and secretion over time in vitro (Fig. 1b).

3.2.8 Days 7 and 14: Chemiluminescent Assay Analysis of SEAP Secretion

1. Transfer 5 μL of serum from each sample per well, in duplicates, to a 96-well microtiter plate (*see* **Note 18**).
2. Prepare a 1× Dilution Buffer by diluting the supplied 5× Dilution Buffer 1:5 with distilled H_2O.
3. Add 37.5 μL of 1× Dilution Buffer to each sample well in the plate.
4. Cover the plate with a lid, wrap the plate using aluminum foil, and incubate the samples for 30 min at 65 °C using a heating block.
5. Place samples on ice for 3 min to cool and then allow to equilibrate to room temperature on the bench top.

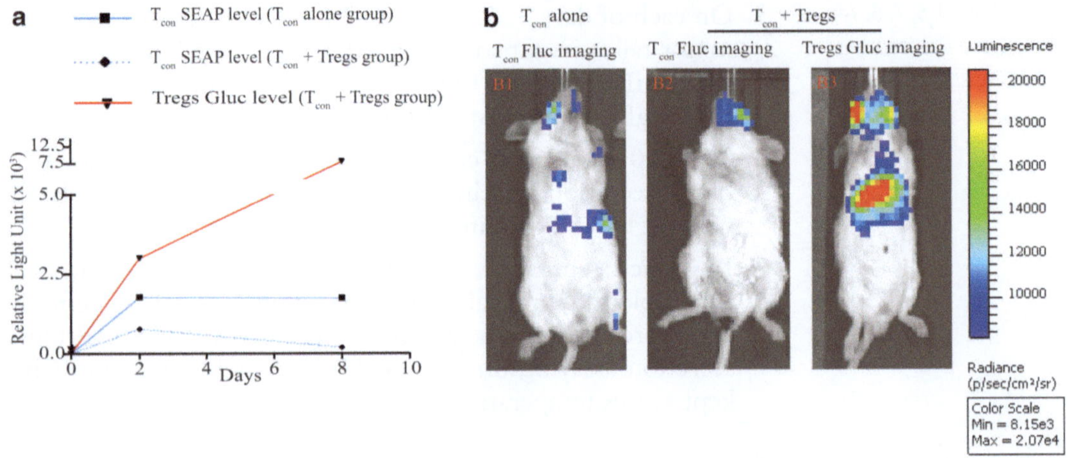

Fig. 2 Irradiated BALB/c mice received either C57BL/6 CD4+ CD25− T_{con} cells alone or in combination with T_{regs}. T_{con} cells were transduced with lentivirus vector encoding Fluc, SEAP and mCherry; T_{regs} were transduced with lentivirus vector encoding Gluc and GFP. (**a**) Gluc and SEAP activity measured in the blood of the BALB/c mice over time. (**b**) Bioluminescent imaging of the BALB/c mice that received either the transduced T_{con} cells either alone (b1), or with transduced T_{regs} (b2 and b3). (b1) and (b2) shows Fluc signal from T_{con} without T_{regs} (b1) or with co-transfer of T_{regs} (b2). Fluc signal is significantly less in the presence of T_{regs} that suppress T_{con} proliferation. (b3) represents Gluc signal from T_{regs} of the same group

6. To each sample add 50 μL of SEAP Substrate Solution and then incubate for 30 min at room temperature.
7. Measure the chemiluminescent activity using a plate luminometer by acquiring photon counts for 10 s.
8. Taken with Subheading 3.2.7, expected results demonstrate the production of Gluc by live proliferating cells increases over time (Fig. 2a); the transfer of T_{con} alone induces an increase in SEAP production, reaching a plateau over time. However, the co-transfer of T_{regs} reduces T_{con} cell proliferation as indicated by a continual decline in the level of SEAP.

3.2.9 Days 3, 7, 14: Multimodal Bioluminescent Mouse Imaging Using Gluc and Fluc Activity

1. Prepare d-Luciferin solution by dissolving stock powder in PBS at a final concentration of 25 mg/mL and keep on ice.
2. Fully anesthetize mice using an isoflurane inhalation chamber system.
3. Immediately before use, at a dose of 4 mg/kg body weight, for each 25 g mouse, mix 20 μL of coelenterazine (5 mg/mL stock prepared in acidified methanol) with 130 μL cold PBS and administer immediately into a mouse via intraocular injection (*see* **Note 19**).
4. Immediately after coelenterazine injection, acquire photon counts using a cooled CCD camera with no illumination (*see* **Notes 20** and **21**).

5. After performing bioluminescent imaging for each mouse, wait at least 10 min to allow for the Gluc signal to be fully eliminated from detectable values (*see* **Note 22**). Perform a test photon acquisition to ensure that only background signal persists.
6. Re-anesthetize mice using isoflurane.
7. Administer 150 μL of d-Luciferin, thawed on ice, per mouse via intraperitoneal injection. As many mice can be injected at the same time as the cooled CCD camera chamber has capacity for.
8. Return mice to anesthesia chamber for 10 min.
9. Acquire photon counts using a cooled CCD camera with no illumination (*see* **Note 20**).
10. For all images captured, on both Gluc and Fluc activity, use the Living Image software to calculate the sum of the photon counts in regions of activity by defining the regions of interest around the tumors using an automatic intensity contour procedure. This will identify the bioluminescent signals with intensities significantly greater than the background.
11. Expected results should show a clear signal detected around the spleen as compared to negative controls (Fig. 1d), showing that this method could be used to localize T_{regs} noninvasively in vivo. Furthermore, if flow cytometry is performed to analyze pancreatic LNs and the pancreas from the NOD-SCID mice expected results reveal a distinct $CD4^{+}GFP^{+}$ population (Fig. 1e), confirming the trafficking of T_{regs} to the pancreas.
12. As expected results, mice that received T_{regs} in addition to T_{con} showed a clearly reduced Fluc signal around the spleen (Fig. 2b2) compared to mice that received only T_{con} (Fig. 2b1). Mice that received T_{regs} and T_{con} show a clear Gluc signal detected around the spleen (Fig. 2b3), showing that this method could be used to also track a second type of T cells simultaneously in vivo.

4 Notes

1. 2× HeBS solution aliquots should be stored long-term at −80 °C but must be at room temperature when used for transfecting; keep one aliquot at room temperature as it is stable for several weeks before affecting transfection efficiency. pH of 7.00 is crucial, as below 6.95 no precipitates will form and above 7.05 precipitates will be too coarse. If the pH is outside of this critical window then there will be very poor transfection efficiency as a result. Also critically important is to use autoclaved distilled or double-distilled H_2O but not MilliQ H_2O.

2. pCSCW-Gluc-IRES-GFP plasmid vector expresses Gluc and the green-fluorescent protein separated by an IRES element, all under the control of the constitutively active cytomegalovirus promoter [18].
3. pCSCW-Fluc-IRES-mCherry plasmid vector expresses Fluc and the mCherry fluorescent protein separated by an IRES element under the CMV promoter [18].
4. pCSCW-SEAP-IRES-mCherry plasmid vector expresses SEAP and the mCherry fluorescent protein separated by an IRES element under the CMV promoter [18].
5. Coelenterazine is the *Gaussia* luciferase substrate. It is critical to prepare a solution of 5 mg/mL stock using acidified methanol (methanol with a drop of hydrochloric acid added) from the powder provided. Aliquot and store at −20 °C. Coelenterazine is light sensitive so do not expose needlessly to light.
6. d-Luciferin is the *Firefly* luciferase substrate. Store at −80 °C and thaw on ice when needed for use. d-Luciferin is light sensitive.
7. Supplement DMEM culture medium for final composition of 10 % fetal bovine serum and 1 % penicillin/streptomycin, mix by shaking, and store at 4 °C.
8. Perform large-scale lentiviral packaging and concentration at least 2 weeks prior to T cell isolation and activation days. Note that each large-scale lentiviral preparation requires approximately 2.1×10^8 293T cells. In this case, three separate constructs were packaged, thus requiring approximately 6.3×10^8 293T cells to be cultured in advance of lentiviral preparation. Also note that the following protocol applies to only one large-scale lentiviral preparation. This must be repeated an additional two times in order to package all three desired plasmid vectors (pCSCW-Gluc-IRES-IGFP, pCSCW-Fluc-IRES-mCherry, and pCSCW-SEAP-IRES-mCherry).
9. Contents of Vial 1 must be mixed drop-wise to Vial 2 while vortexing, likewise, make sure to cap Vial 2 without ceasing vortexing and then increase vortex to max speed for 1 full min. Adding mixtures too rapidly, and any cessation of vortexing during this time, may result in failed or low-yield transfection.
10. Supplement Opti-MEM culture medium for final composition of 1 % penicillin/streptomycin and mix by shaking, store at 4 °C.
11. Prepare sterile MACS buffer for a final composition of 0.5 % BSA, 2 mM EDTA in sterile PBS. Do so by adding 2.5 g BSA and 2 mL of sterile 0.5 M EDTA to 500 mL of sterile PBS and filter through a Nalgene Rapid-Flow filter unit. Maintain on ice until required; store thereafter at 4 °C.

12. Prepare sterile RPMI-1640 complete culture medium for a final composition of 10 % newborn calf serum, 1 % penicillin/streptomycin, and 1 %l-Glutamine. Mix by shaking. Maintain on ice until required; store thereafter at 4 °C.
13. Calculate total volumes needed of each master mix, where each well containing 5×10^4 activated cells to be infected requires 1 μL of each respective virus, 1 μL of 1:1,000 diluted polybrene for a final concentration of 0.8 μg/mL polybrene, and volume up to 10 μL with supplemented RPMI-1640 medium; calculate for extra wells to account for standard pipetting error.
14. For bone marrow collection, the femur should be flushed from both sides of the bone marrow canal while flushing from only one side proves to be sufficient for the tibia.
15. Retro-orbital administration is to be performed by appropriately trained individuals only as piercing of the eye can lead to blindness, and overly rapid administration can lead to mouse death. Mice are anesthetized with isoflurane. Using an insulin syringe, cells will be transferred in a single cell suspension not exceeding 200 μL into the venous sinus at the outer canthus of the eye.
16. Setting the pipettor to 45 μL and first aspirating up the 5 μL of 50 mM EDTA into the pipette tip to prewet the tip prior to collecting the 40 μL of whole blood is a simple and effective way to significantly decrease chances of blood clotting forming during blood collection. Be sure to mix thoroughly.
17. Note that the plate must be a opaque white 96-well plate for assaying. Clear plates cause cross-contamination of bioluminescent signal and must not be used.
18. Note that the following protocol utilizes the Great EscAPe SEAP Chemiluminescence Kit. Allow the supplied SEAP Substrate Solution to equilibrate to room temperature prior to beginning assay protocol. Also note that there should be more than 10 μL of stored serum per mouse per collection day: each sample only requires 5 μL of serum, thus duplicate samples may be assayed. Again only opaque white 96-well plates should be used for assaying.
19. Note that due to the kinetics of Gluc, it is optimal to inject coelenterazine into and image only one mouse at a time. Also note to keep coelenterazine and PBS on ice before use otherwise large precipitates may form during mixing. A small precipitate/cloudy brown solution might form during injection which normally does not interfere with imaging. However, if coelenterazine diluted in PBS is incubated for a long period of time, a larger precipitate will form which might lead to blood clots and mouse death upon injection.

20. Typically, use of the Living Image software's automatic exposure time will yield the best result, though one can manually acquire over a time of 30–60 s.
21. Note one might observe some photons around the ocular area which can be due to the way the injection was performed, however this is not usually detrimental to the experiment or the mouse.
22. Note that during this 10 min recovery time the mice do not need to be under anesthesia.

References

1. Sakaguchi S, Sakaguchi N, Asano M, Itoh M, Toda M (1995) Immunologic self-tolerance maintained by activated T cells expressing IL-2 receptor alpha-chains (CD25). Breakdown of a single mechanism of self-tolerance causes various autoimmune diseases. J Immunol 155(3): 1151–1164
2. Itoh M et al (1999) Thymus and autoimmunity: production of CD25+CD4+ naturally anergic and suppressive T cells as a key function of the thymus in maintaining immunologic self-tolerance. J Immunol 162(9):5317–5326
3. Chen W et al (2003) Conversion of peripheral CD4+CD25- naive T cells to CD4+CD25+ regulatory T cells by TGF-beta induction of transcription factor Foxp3. J Exp Med 198(12): 1875–1886
4. Curotto de Lafaille MA et al (2008) Adaptive Foxp3+ regulatory T cell-dependent and -independent control of allergic inflammation. Immunity 29(1):114–126
5. Nishimura E, Sakihama T, Setoguchi R, Tanaka K, Sakaguchi S (2004) Induction of antigen-specific immunologic tolerance by in vivo and in vitro antigen-specific expansion of naturally arising Foxp3+CD25+CD4+ regulatory T cells. Int Immunol 16(8):1189–1201
6. Sakaguchi S (2005) Naturally arising Foxp3-expressing CD25+CD4+ regulatory T cells in immunological tolerance to self and non-self. Nat Immunol 6(4):345–352
7. Collison LW et al (2007) The inhibitory cytokine IL-35 contributes to regulatory T-cell function. Nature 450(7169):566–569
8. Pandiyan P, Zheng L, Ishihara S, Reed J, Lenardo MJ (2007) CD4+CD25+Foxp3+ regulatory T cells induce cytokine deprivation-mediated apoptosis of effector CD4+ T cells. Nat Immunol 8(12):1353–1362
9. Campbell DJ, Ziegler SF (2007) FOXP3 modifies the phenotypic and functional properties of regulatory T cells. Nat Rev Immunol 7(4):305–310
10. Riley JL, June CH, Blazar BR (2009) Human T regulatory cell therapy: take a billion or so and call me in the morning. Immunity 30(5): 656–665
11. Badr CE, Hewett JW, Breakefield XO, Tannous BA (2007) A highly sensitive assay for monitoring the secretory pathway and ER stress. PLoS One 2(6):e571
12. Bao R et al (2002) Activation of cancer-specific gene expression by the survivin promoter. J Natl Cancer Inst 94(7):522–528
13. Bao R, Selvakumaran M, Hamilton TC (2000) Use of a surrogate marker (human secreted alkaline phosphatase) to monitor in vivo tumor growth and anticancer drug efficacy in ovarian cancer xenografts. Gynecol Oncol 78(3 Pt 1): 373–379
14. Hewett JW et al (2007) Mutant torsinA interferes with protein processing through the secretory pathway in DYT1 dystonia cells. Proc Natl Acad Sci USA 104(17):7271–7276
15. Hiramatsu N et al (2006) Secreted protein-based reporter systems for monitoring inflammatory events: critical interference by endoplasmic reticulum stress. J Immunol Methods 315(1–2): 202–207
16. Meng Y et al (2005) Real-time monitoring of mesangial cell-macrophage cross-talk using SEAP in vitro and ex vivo. Kidney Int 68(2): 886–893
17. Meng Y et al (2005) Continuous, noninvasive monitoring of local microscopic inflammation using a genetically engineered cell-based biosensor. Lab Invest 85(11):1429–1439
18. Wurdinger T et al (2008) A secreted luciferase for ex vivo monitoring of in vivo processes. Nat Methods 5(2):171–173
19. Contag CH, Ross BD (2002) It's not just about anatomy: in vivo bioluminescence imaging as an eyepiece into biology. J Magn Reson Imaging 16(4):378–387
20. Weissleder R, Ntziachristos V (2003) Shedding light onto live molecular targets. Nat Med 9(1): 123–128
21. Berger J, Hauber J, Hauber R, Geiger R, Cullen BR (1988) Secreted placental alkaline

phosphatase: a powerful new quantitative indicator of gene expression in eukaryotic cells. Gene 66(1):1–10

22. Cullen BR, Malim MH (1992) Secreted placental alkaline phosphatase as a eukaryotic reporter gene. Methods Enzymol 216: 362–368
23. Hiramatsu N, Kasai A, Hayakawa K, Yao J, Kitamura M (2006) Real-time detection and continuous monitoring of ER stress in vitro and in vivo by ES-TRAP: evidence for systemic, transient ER stress during endotoxemia. Nucleic Acids Res 34(13):e93
24. Hiramatsu N et al (2005) Alkaline phosphatase vs luciferase as secreted reporter molecules in vivo. Anal Biochem 339(2):249–256
25. Shiraiwa T et al (2007) Establishment of a non-invasive mouse reporter model for monitoring in vivo pdx-1 promoter activity. Biochem Biophys Res Commun 361(3): 739–744
26. Abruzzese RV et al (1999) Ligand-dependent regulation of plasmid-based transgene expression in vivo. Hum Gene Ther 10(9): 1499–1507
27. Chastain M et al (2001) Antigen levels and antibody titers after DNA vaccination. J Pharm Sci 90(4):474–484
28. Muller L, Saydam O, Saeki Y, Heid I, Fraefel C (2005) Gene transfer into hepatocytes mediated by herpes simplex virus-Epstein-Barr virus hybrid amplicons. J Virol Methods 123(1): 65–72
29. Phuong LK et al (2003) Use of a vaccine strain of measles virus genetically engineered to produce carcinoembryonic antigen as a novel therapeutic agent against glioblastoma multiforme. Cancer Res 63(10):2462–2469
30. Hewett JW et al (2008) siRNA knock-down of mutant torsinA restores processing through secretory pathway in DYT1 dystonia cells. Hum Mol Genet 17(10):1436–1445
31. Tannous BA, Kim DE, Fernandez JL, Weissleder R, Breakefield XO (2005) Codon-optimized Gaussia luciferase cDNA for mammalian gene expression in culture and in vivo. Mol Ther 11(3):435–443
32. Venisnik KM, Olafsen T, Gambhir SS, Wu AM (2007) Fusion of Gaussia luciferase to an engineered anti-carcinoembryonic antigen (CEA) antibody for in vivo optical imaging. Mol Imaging Biol 9(5):267–277
33. Zhou L, Chong MM, Littman DR (2009) Plasticity of CD4+ T cell lineage differentiation. Immunity 30(5):646–655
34. Tannous BA (2009) Gaussia luciferase reporter assay for monitoring biological processes in culture and in vivo. Nat Protoc 4(4):582–591
35. Ablamunits V, Elias D, Cohen IR (1999) The pathogenicity of islet-infiltrating lymphocytes in the non-obese diabetic (NOD) mouse. Clin Exp Immunol 115(2):260–267

Part V

Emerging Technologies

Chapter 18

Multiplex Functional Bioluminescent Reporters Using Gaussia Luciferase Fused to Epitope Tags in an Immunobinding Assay

Sjoerd van Rijn, Thomas Würdinger, and Jonas Nilsson

Abstract

The use of *Gaussia* luciferase in a multiplex assay can have several advantages over the singleplex method for an experimental setup. Issues such as intersample variability, screening purposes, efficiency, and in vivo applications can be addressed using a multiplex assay. Here we describe a functional reporter multiplex method using *Gaussia* luciferase fused to epitope tags to identify the different reporters that are expressed. Tag specific antibodies are used to bind and separate the tagged luciferase reporters.

Key words *Gaussia* luciferase, Optical imaging, Bioluminescent imaging, Multiplex assay, Epitope tag, Antibody binding assay

1 Introduction

Bioluminescent reporters can be used to noninvasively detect and quantify biological processes in vitro and in vivo [1]. By engineering the reporter gene under the control of genetic promoters or other gene regulatory elements, the resulting reporter protein reflects the regulation of these elements. Examples of these regulatory elements can be constitutively active promoters to analyze cell survival and proliferation (e.g., CMV promoter) or specific promoters containing transcription factor regulatory elements (TREs) to analyze transcription factor activities [2, 3]. Other examples are 3′-UTR regions to analyze miRNA activities [4] and protease cleavage sites (PCSs) to analyze protease activities [5]. Regularly used bioluminescent reporter genes are *Firefly* luciferase [6] (Fluc) and *Renilla* luciferase [7] (Rluc). After translation these luciferases are localized inside the cell's cytoplasm and ex vivo detection can therefore be a challenge. The enzyme *Gaussia* luciferase [8] (Gluc) is secreted into the extracellular fluids such as culture medium,

Christian E. Badr (ed.), *Bioluminescent Imaging: Methods and Protocols*, Methods in Molecular Biology, vol. 1098,
DOI 10.1007/978-1-62703-718-1_18,

Fig. 1 Overview of the $Gluc_{Tag}$ construct. We constructed ten $Gluc_{Tag}$ constructs, each with a different tag: $Gluc_{Flag}$, $Gluc_{His}$, $Gluc_{HA}$, $Gluc_{AcV5}$, $Gluc_{V5}$ and $Gluc_{Glu}$, $Gluc_{Myc}$, $Gluc_{Kt3}$, $Gluc_{Au1}$, and $Gluc_{E2}$

animal blood, and urine and can therefore be readily noninvasively detected in these fluids enabling serial sampling over time in a single experiment.

One major disadvantage of luciferase reporters is the difficulty to use a multiplicity of luciferase reporters in combination in a single experiment. This limits the use of luciferase reporters in multiplex assays. Attempts are made to use a range of reporters using modified luciferases, each producing light with a different wavelength in order to discriminate between the different parameters they represent. The downside of this strategy is the need for sophisticated equipment to separate the different signals. Also, overlap in the light spectra might make signal separation a challenge and the number of reporters limited.

Another strategy to analyze multiple reporters in a single assay is to use luciferases that need different enzyme substrates to produce light [9]. A regularly used example is *firefly* luciferase, which needs D-luciferin as a substrate, in combination with *Renilla* luciferase, which needs coelenterazine. The disadvantage of this method is the limited number of luciferase substrate combinations currently known.

We aimed to develop an easily adaptable method for multiplexing a secreted luciferase reporter in order to perform broad range screening of biologically relevant processes in vitro or in vivo. Therefore, we engineered a range of fusion gene reporters consisting of the Gluc gene, each coupled to a different epitope tag [10] (*see* Fig. 1). By this method we engineered the ten $Gluc_{Tag}$ reporters $Gluc_{Flag}$, $Gluc_{His}$, $Gluc_{HA}$, $Gluc_{AcV5}$, $Gluc_{V5}$ and $Gluc_{Glu}$, $Gluc_{Myc}$, $Gluc_{Kt3}$, $Gluc_{Au1}$, and $Gluc_{E2}$. Epitope tags [11] are peptide sequences usually derived from viral or bacterial genes, giving them a high affinity to mammalian antibodies. Coupling the Gluc gene to a range of different epitope tags enables us to design a large number of different $Gluc_{Tag}$ reporters. Using the corresponding tag specific antibodies to selectively immunobind the different $Gluc_{Tag}$ makes it possible to separate the different $Gluc_{Tag}$ reporters and individually quantify them by reading out the light production after addition of substrate [10] (*see* Fig. 3). As a proof of concept we designed ten different $Gluc_{Tag}$ based reporters, each coupled to a different epitope tag, under the control of the constitutively active cytomegalovirus (CMV) promoter. After lentiviral transduction [12] we produced ten U87 glioma cell lines, each stably expressing one $Gluc_{Tag}$ reporter and used these cells to collect $Gluc_{Tag}$ conditioned cell culture medium or mouse blood to monitor in vitro and in vivo tumor cell growth by measuring $Gluc_{Tag}$ activity ex situ, after tag specific antibody immunobinding (*see* Fig. 3). In this chapter the proof of concept of the multiplex Gaussia luciferase-based functional reporter assays in vivo is described [10].

2 Materials

2.1 Construction and Development of $Gluc_{Tag}$ Reporter Cell Lines

1. The CSCW-Gluc-IRES-CFP lentiviral vector DNA (*see* **Note 1**). This lentiviral vector co-expresses the Gluc bioluminescent reporter and the cerulean fluorescent protein (CFP) control. The internal ribosomal entry site (IRES) allows co-expression of both proteins using the same cytomegalovirus (CMV) promoter. The vector is designated CSCW-Gluc-CFP in this protocol.
2. AccuPrime Pfx DNA polymerase (Life Technologies, *see* **Note 2**). It is provided with a 10× AccuPrime reaction mix and separate $MgSO_4$ (50 mM).
3. PCR amplification primer oligonucleotides (10 μM, *see* **Note 1**):
   (a) Forward: 5′-GCG TGT ACG GTG GGA GGT CT-3′.
   (b) Reverse: 5′-T CTC GAG TAG TCT AGA GTC ACC ACC GGC CCC CTT-3′.
4. PCR thermocycler.
5. Restriction enzymes NheI, XbaI, and XhoI (*see* **Note 3**).
6. Heat block.
7. DNA gel electrophoresis system.
8. 0.5 % agarose (v/w) DNA gel with ethidium bromide (10 μg/ml final concentration). The total volume depends on the DNA gel electrophoresis system used in (**item 7**). Add for example 0.5 g agarose to 100 ml TAE buffer (**item 9**) and heat it in a microwave for 2 min at the highest power (~900 W). Make sure that the agarose has melted completely. Cool down the gel to ~65 °C, add ethidium bromide (10 μg/ml final concentration, *see* **Note 4**) and swirl to mix. Pour gel in a DNA gel cast with a DNA gel slot comb and let it solidify for 30 min. Remove the DNA gel slot comb carefully.
9. 1× TAE buffer (40 mM Tris, 20 mM acetic acid and 1 mM EDTA in dH_2O). Make a 50× TAE stock solution by dissolving 242 g Tris base in dH_2O. Add 57.1 ml glacial acetic acid and then add 100 ml 0.5 M EDTA (pH 8.0). Add the volume up to 900 ml with dH_2O and set the pH to 8.5 using HCl or NaOH. Bring the total volume to 1,000 ml with distilled H_2O and autoclave the TAE buffer for 20 min at 121 °C. Dilute 20 ml 50× TAE stock solution in 980 ml distilled H_2O to prepare 1,000 ml 1× TAE buffer.

10. DNA gel loading dye and a broad range (100 bp to 10,000 bp) DNA gel molecular weight marker. Use the loading dye and the weight marker according to the manufacturer's guidelines.
11. DNA gel extraction kit.
12. T4 DNA ligase.
13. XL10-Gold ultracompetent cells (Agilent Technologies) for large or ligated DNA.
14. LB + ampicillin (50 μg/ml) bacteria selection and propagation plates. To make LB, measure 10 g of peptone, 5 g yeast extract, and 10 g NaCl and suspend in 1,000 ml distilled H_2O and autoclave the suspension for 20 min at 121 °C. Let the LB cool down to ~65 °C and add ampicillin to a final concentration of 50 μg/ml and briefly swirl. Pour 20 ml of LB + ampicillin per plate (100 mm diameter).
15. 37 °C bacteria incubator.
16. QIAfilter plasmid kit.
17. Epitope tag sense and antisense oligonucleotides with appropriate ligation overhangs (*see* Table 1 and **Note 1**) in a concentration of 100 μM.
18. 10× Oligonucleotide annealing buffer. To prepare a 50 ml stock solution, mix 5 ml 1 M Tris–HCl (pH 7.5, final concentration 100 mM) with 10 ml 5 M NaCl (final concentration 1 M) and 1 ml 0.5 M EDTA (final concentration 10 mM) and add 34 ml ultrapure H_2O.
19. Human embryonic kidney cell line HEK293T.
20. Human glioblastoma cell line U87.
21. Dulbecco's modified Eagle's medium (DMEM) containing 4.5 g/L glucose, 110 mg/L sodium pyruvate, and 584 mg/L L-glutamine. To make DMEM complete culture medium, add 10 % Fetal bovine serum and penicillin/streptomycin.
22. Phosphate buffered saline (PBS).
23. Trypsin (0.5 mg/ml) + EDTA (0.22 mg/ml) in PBS.
24. A cell culture incubator at 37 °C and with 5 % CO_2.
25. Third-generation lentiviral packaging plasmid pRRE, packaging plasmid pRSV/Rev, and envelope plasmid pMD2.G (Addgene).
26. 3 M Calcium chloride ($CaCl_2$) to make 100 ml of 3 M $CaCl_2$ solution, dissolve 33.3 g $CaCl_2 \times 2H_2O$ in 95 ml ultrapure H_2O. Set the pH to 7.2 and fill the solution up to 100 ml. Sterile filter the 3 M $CaCl_2$ solution.
27. 2× HEPES buffered saline (HEBS). Dissolve 8 g NaCl (final concentration 273.7 mM), 0.37 g KCl (final concentration 9.9 mM), 1 g Na_2HPO_4 (1.4 mM), 1 g dextrose (final concentration 11.1 mM), and 5 g HEPES (42 mM) in 450 ml ultrapure H_2O. Set the pH to 7.2 and then fill up the solution to 500 ml with ultrapure H_2O. Autoclave the 2× HEBS for 15 min at 121 °C.

Table 1
Epitope sequences and antibodies used in the immunobinding assays and immunostainings. The epitope tag oligonucleotides are designed to complement each other and to create the proper ligation overhang (XbaI-tag sequence-XhoI). The assay is optimized for use with epitope tag Flag, His, HA, AcV5, V5, and Glu. They show a high binding and assay performance in vitro and in vivo. Epitope tag combinations Myc, Kt3, Au1, and E2 show a low binding and assay performance due to a lower antibody affinity. Conditions for use of these epitope tags or other tags in the assay need to be optimized

Tag	Oligonucleotide sequence: sense, antisense (5′–3′)	Antibody
Flag	CTAGAGACTACAAAGACCATGACGGTGATTATAAAGATCATGACATCGACTACAAGGATGACGATGACAAGTGAC	Sigma-Aldrich
	TCGAGTCACTTGTCATCGTCATCCTTGTAGTCGATGTCATGATCTTTATAATCACCGTCATGGTCTTTGTAGTCT	F1804
His	CTAGACATCATCACCATCACCACTGAC	Abcam
	TCGAGTCAGTGGTGATGGTGATGATGT	ab81663
HA	CTAGATATCCGTATGATGTGCCGGATTATGCGTGAC	Abcam
	TCGAGTCACGCATAATCCGGCACATCATACGGATAT	ab59076
AcV5	CTAGAAGCTGGAAGGACGCCAGCGGCTGGAGCTGAC	Abcam
	TCGAGTCAGCTCCAGCCGCTGGCGTCCTTCCAGCTT	ab49581
V5	CTAGAGGCAAGCCTATCCCTAACCCTCTGCTGGGCCTGGACAGCACCTGAC	Abcam
	TCGAGTCAGGTGCTGTCCAGGCCCAGCAGAGGGTTAGGGATAGGCTTGCCT	ab27671
Glu	CTAGATGCGAGGAAGAGGAATACATGCCTATGGAGTGAC	Abcam
	TCGAGTCACTCCATAGGCATGTATTCCTCTTCCTCGCAT	ab24627
Myc	CTAGAGAACAAAAACTCATCTCAGAAGAGGATCTGTGAC	Sigma-Aldrich
	TCGAGTCACAGATCCTCTTCTGAGATGAGTTTTTGTTCT	M4439
Kt3	CTAGAAAGCCTCCAACACCTCCACCTGAGCCTGAGACCTGAC	Abcam
	TCGAGTCAGGTCTCAGGCTCAGGTGGAGGTGTTGGAGGCTTT	ab24739
Au1	CTAGAGACACCTACAGATACATCTGAC	Covance
	TCGAGTCAGATGTATCTGTAGGTGTCT	MMS-130P
E2	CTAGAAGCAGCACCAGCAGCGACTTCAGAGACAGATGAC	Abcam
	TCGAGTCATCTGTCTCTGAAGTCGCTGCTGGTGCTGCTT	ab977

2.2 $Gluc_{Tag}$-CFP Multiplex Assay Application

2.2.1 $Gluc_{Tag}$ Immunobinding Assay In Vitro

1. 6-well cell culture plates.
2. A cell culture incubator at 37 °C and with 5 % CO_2.
3. Wash buffer. Add 0.5 ml Tween-20 (0.05 %) to 1 L of phosphate buffered saline (PBS) to make the PBS + 0.05 % Tween-20 wash buffer solution.
4. White goat anti-mouse IgG coated 96-well microplates (Thermo Scientific, *see* **Note 5**).
5. Mouse monoclonal anti-Tag IgG (*see* Table 1, *see* **Note 6**).
6. Microplate centrifuge.
7. Microplate shaker.
8. Block buffer. Add 0.5 ml Tween-20 (0.05 %) and 10 g bovine serum albumin (BSA, 1 % v/w) in 1 L of phosphate buffered saline (PBS) to make the PBS + 0.05 % Tween-20 + 1 % BSA block buffer solution.
9. Coelenterazine Gluc substrate. Dilute coelenterazine in methanol (5 mg/ml) to make a stock solution. Before use, dilute the stock solution coelenterazine 1 μl per 1,000 μl PBS + Triton X-100. Incubate the final user solution coelenterazine (5 μg/ml) 30 min at room temperature.
10. Microplate luminometer.

2.2.2 $Gluc_{Tag}$ Immunostaining In Vitro

1. 24-well plates.
2. Phosphate buffered saline (PBS).
3. 3.7 % formaldehyde fixative solution.
4. Block buffer. Add 10 g bovine serum albumin (BSA) to 1 L of PBS to make a PBS + 1% BSA block buffer solution.
5. Permeabilization buffer. Add 1 ml Triton X-100 to 1 L of PBS to make a PBS + 0.1% Triton X-100 permeabilization buffer solution.
6. Mouse monoclonal anti-Tag IgG (*see* Table 1, *see* **Note 6**).
7. Goat polyclonal anti-mouse-HRP Ig (Dako).
8. DAB substrate kit.
9. Light microscope.

2.2.3 $Gluc_{Tag}$ Immunobinding Assay In Vivo

1. Athymic nude-foxn1^{nu} mice.
2. Isoflurane anesthesia system.
3. Temgesic (Buprenorphine hydrochloride) analgesia in PBS.
4. Small animal stereotaxic frame.
5. Surgical scalpel.
6. Microsyringe (Hamilton).
7. Microdrill 0.8 mm.

8. Capillary collection and sample container (Sarstedt, *see* **Note** 7).
9. Blood dilution buffer. Add 0.8 g EDTA (0.8 % v/w) to 100 ml PBS to make PBS + 0.8 % EDTA blood dilution buffer solution.
10. White goat anti-mouse IgG coated 96-well microplates (Thermo Scientific, *see* **Note 5**).
11. Wash buffer. Add 0.5 ml Tween-20 (0.05 %) to 1 L of phosphate buffered saline (PBS) to make the PBS + 0.05 % Tween-20 wash buffer solution.
12. Mouse monoclonal anti-Tag IgG (*see* Table 1, *see* **Note 6**).
13. Microplate centrifuge.
14. Block buffer. Add 0.5 ml Tween-20 (0.05 %) and 10 g bovine serum albumin (BSA, 1 % v/w) in 1 L of phosphate buffered saline (PBS) to make the PBS + 0.05 % Tween-20 + 1 % BSA block buffer solution.
15. Microplate shaker.
16. Coelenterazine Gluc substrate. Dilute coelenterazine in methanol (5 mg/ml) to make a stock solution. Before use, dilute the stock solution coelenterazine 1 μl per 1,000 μl PBS + Triton X-100. Incubate the final user solution coelenterazine (5 μg/ml) 30 min at room temperature.
17. Microplate luminometer.

2.2.4 $Gluc_{Tag}$ Immunostaining In Vivo

1. Formaldehyde 3.7 % in PBS. To make 100 ml of 3.7 % formaldehyde in PBS, add 10 ml of 37 % stock solution formaldehyde to 90 ml of phosphate buffered saline (PBS) and mix.
2. Microtome.
3. Glass microscope slides.
4. Xylene.
5. Ethanol series. Make three ethanol solutions of 100 %, 97 % and 75 % ethanol in H_2O. To make 100 ml of 97 % ethanol solution, add 97 ml ethanol in 3 ml H_2O and mix. To make 100 ml of 75 % ethanol solution, add 75 ml ethanol to 25 ml H_2O and mix.
6. 0.3 % H_2O_2 in methanol. To make a 0.3 % hydrogen peroxide (H_2O_2) in methanol solution, add 1 ml of 30 % H_2O_2 stock solution to 100 ml of methanol and mix.
7. Antigen retrieval citrate buffer. To make the antigen retrieval citrate buffer, add 1.92 g anhydrous citric acid (final concentration 10 mM) to 950 ml of H_2O and mix. Set the pH to 6 and fill up the solution to 1 L. Finally, add 0.5 ml of Tween-20 and mix well.
8. Microwave.

9. Mouse monoclonal anti-Tag IgG (*see* Table 1, *see* **Note 6**).
10. Antibody diluent (Dako).
11. Phosphate buffered saline (PBS).
12. Goat polyclonal anti-mouse-HRP Ig (Dako).
13. DAB substrate kit.
14. Hematoxylin staining solution.
15. Light microscope.

3 Methods

3.1 Construction and Development of $Gluc_{Tag}$ Reporter Cell Lines

1. Construct the $Gluc_{modified}$ gene by amplifying the $Gluc_{parental}$ gene from lentiviral vector CSCW-Gluc-CFP (*see* Fig. 2 and **Note 1**) making a PCR reaction mix of 20 μl using primers that exclude the STOP codon and add the unique restriction site XbaI downstream. Use a high fidelity proof reading DNA polymerase and follow the manufacturer's guidelines to amplify the $Gluc_{modified}$ gene (633 bp) in thermal cycler by (1) denature the plasmid DNA for 2 min at 95 °C. Then (2) denature for 30 s at 95 °C, (3) anneal primers for 30 s at 62 °C and (4) elongate DNA for 1 min at 72 °C and cycle sequence 2, 3, 4 for 35 times. Allow (5) final elongation of DNA for 2 min at 72 °C.
2. To form sticky ends to the $Gluc_{modified}$ gene in order to clone it back into the CSCW lentiviral backbone to replace the $Gluc_{parental}$ gene, restrict the $Gluc_{modified}$ gene with restriction enzyme NheI upstream, and restriction enzyme XhoI downstream. Restrict the total volume of the PCR reaction from **step 1** by making a double digestion using high fidelity enzymes (*see* **Note 3**), according to the manufacturer's guidelines in a total volume of 50 μl. Incubate the restriction mix for 2 h at 37 °C to ensure complete restriction. Also, create the CSCW lentiviral backbone by restricting 2 μg of the parental CSCW-Gluc-CFP lentiviral vector with NheI and XhoI as described to restrict out the $Gluc_{parental}$ gene.
3. Isolate and purify the $Gluc_{modified}$ gene from **step 2** and the CSCW lentiviral backbone from **step 2** on a 0.5 % (w/v) agarose DNA gel. Add DNA loading buffer (final concentration is 1× DNA loading buffer) to the $Gluc_{modified}$ gene and CSCW lentiviral backbone restriction mixtures from **step 2** and load both the samples into a separate gel well. Also load a separate well with a DNA molecular weight marker to identify the product and confirm product sizes. Run the gel in 1× TAE buffer at 100 V until the $Gluc_{parental}$ gene (718 bp) has separated properly from the CSCW lentiviral backbone (9,408 bp).

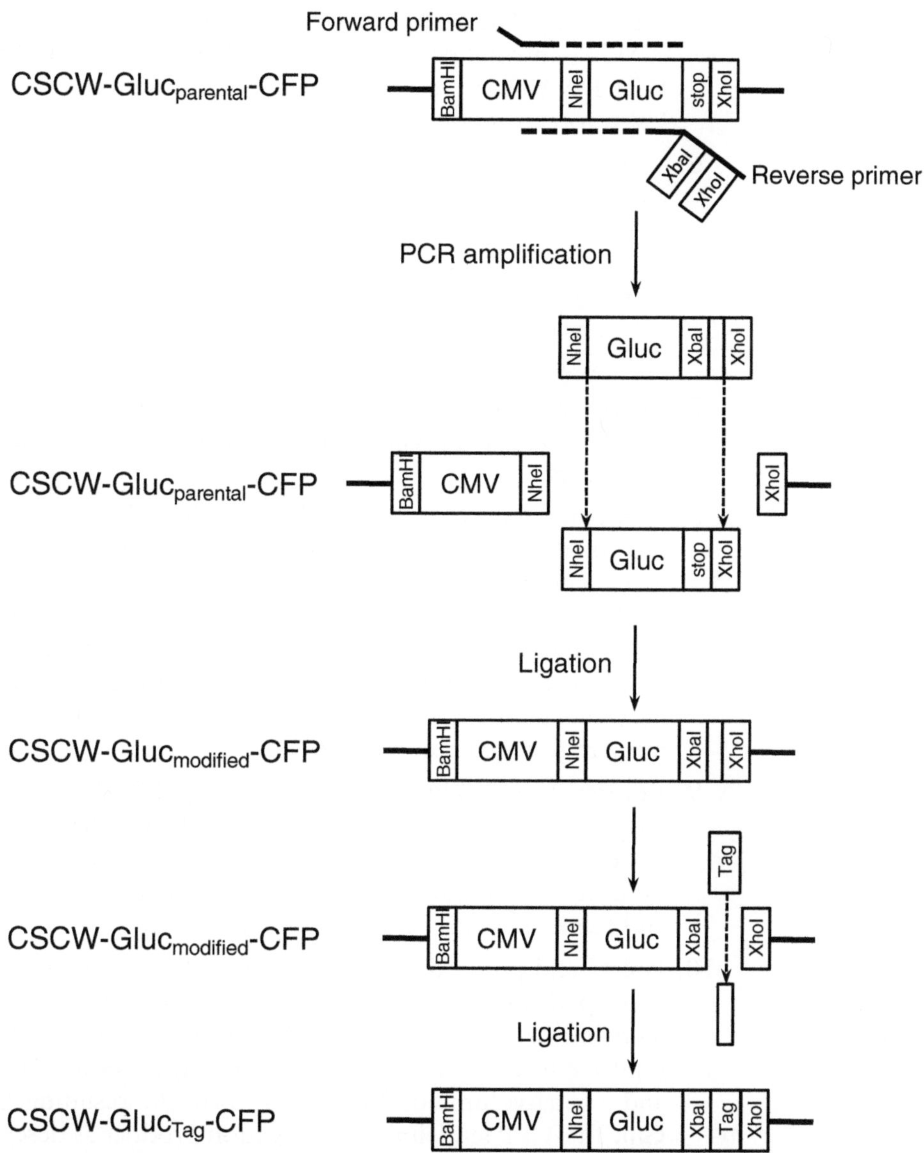

Fig. 2 $Gluc_{Tag}$ cloning strategy. The Gluc gene is amplified from the parental vector using primers that amplify Gluc without the TATA-box and adding an XbaI and XhoI restriction site downstream. This modified Gluc gene then replaces the original Gluc gene in the parental vector to construct the $Gluc_{modified}$ vector. Then, the vector is restricted with XbaI and XhoI to insert an epitope tag of choice. This $Gluc_{Tag}$ is the final construct

4. Extract the $Gluc_{modified}$ gene construct (570 bp) and the CSCW lentiviral backbone (9,408 bp) using a DNA gel extraction kit according to manufacturer's protocol. After elution with 50 μl elution buffer, heat both the DNA extractions in an open tube to 60 °C for 5 min to completely evaporate any residual ethanol.
5. To clone the $Gluc_{modified}$ gene into the CSCW lentiviral backbone, ligate 100 ng CSCW lentiviral backbone from **step 4**

with 57 ng $Gluc_{modified}$ gene from **step 4** (($Gluc_{modified}$ gene): (CSCW lentiviral backbone) is 10:1 M ratio, *see* **Note 8**) using a T4 DNA ligase according to the manufacturer's guidelines. Use a total reaction volume of 20 μl.

6. Transform the resulting ligation product CSCW-$Gluc_{modified}$-CFP lentiviral vector in ultracompetent bacterial cells for ligated DNA according to the manufacturer's guidelines in order to make bacterial clones for CSCW-$Gluc_{modified}$-CFP lentiviral vector production. Plate the bacterial cells on a LB + ampicillin (50 μg/ml) bacteria selection and propagation plate and grow the bacterial colonies overnight at 37 °C.
7. Amplify and isolate the CSCW-$Gluc_{modified}$-CFP lentiviral vector using a DNA plasmid kit following the manufacturer's guidelines.
8. Construct epitope tag inserts by annealing the corresponding epitope tag sense and antisense oligonucleotides (*see* Table 1). Mix 50 μl of H_2O with 10 μl of the 10× annealing buffer, 20 μl of the sense oligonucleotide (100 μM) and 20 μl of the antisense oligonucleotide (100 μM). Heat the annealing mixture in a heat block for 10 min at 65 °C, then take out the metal heat block insert to very slowly cool down the annealing mixture to room temperature.
9. Restrict 2 μg of the CSCW-$Gluc_{modified}$-CFP using restriction enzymes XbaI and XhoI to open up the CSCW lentiviral backbone in order to insert the epitope tag from **step 8**. Make a 20 μl double digestion using high fidelity restriction enzymes and follow manufacturer's guidelines.
10. Isolate and purify the CSCW lentiviral backbone as described in **step 3** and clone the epitope tag insert into the CSCW lentiviral backbone as described in **step 5**. Use a molar ratio (epitope insert):(CSCW lentiviral vector, *see* **Note 8**) of 10:1 and a total volume of 20 μl. Transform the resulting CSCW-GlucTag-CFP lentiviral vector ligation product as described in **step 6** and amplify the vector as described in **step 7**.
11. HEK293T and U87 cells are cultured in DMEM complete culture medium at 37 °C and 5 % CO_2. The cells are diluted 1/10 when the culture vessel is ~90 % confluent. To dilute, aspirate the culture medium and was attached cells with PBS. Shake culture vessel gently and aspirate PBS. Detach the cells by adding 1× Trypsin + EDTA, enough to just cover the surface of the culture vessel. Incubate the cells 5 min at 37 °C for 5 min. Resuspend the cells in DMEM complete culture medium (>5× the Trypsin + EDTA volume) and aspirate 9/10 of the total volume to discard or collect the suspended cells. Add DMEM complete culture medium to the 1/10 leftover

cell suspension up to the final culture volume. Make sure the surface of the culture vessel is covered with ~0.5 cm culture medium. Use room temperature reagents.

12. Produce the lentiviral particles as described in [12]. Transiently transfect 5.5×10^6 HEK293T cells in a 10 cm^2 culture dish with 3 μg pMD2.G envelope plasmid, 5 μg pMDLg/pRRE, and 2.5 μg pRSV/Rev packaging plasmids and 10 μg CSCW-GlucTag-CFP lentiviral vector using the calcium phosphate transfection method (*see* **Note 9**). Harvest the virus-containing medium and spin at $1,000 \times g$ for 5 min to remove residual cells and debris. Aliquot and store the virus-containing medium at 4 °C for use within a week or at −80 °C for longer periods of time.
13. Make U87 cell lines (*see* **Note 10**) stably expressing $Gluc_{Tag}$-CFP by lentiviral transduction as described in [12]. Transduce 2×10^5 U87 cells in a 6 well plate with CSCW-GlucTag-CFP lentivirus (MOI of 5, *see* **Note 11**) overnight. The following day, replace the virus-containing medium with 2 ml fresh DMEM complete culture medium. Expand and culture the cells as described in **step 11**. Transduction efficiency can be determined by fluorescence microscopy of CFP (455–480 nm).

3.2 $Gluc_{Tag}$-CFP Multiplex Assay Application

3.2.1 $Gluc_{Tag}$ Immunobinding Assay In Vitro

1. Plate U87-$Gluc_{6x\ Tag}$-CFP cells (*see* **Note 10**) in a 6-well plate in DMEM complete culture medium and culture the cells in 37 °C and 5 % CO_2. For experimental triplicates, plate 3 wells of a 6-well plate.
2. At predetermined time points, collect 180 μl (30 μl × 6 Tags) of the $Gluc_{6x\ Tag}$ conditioned culture medium of the 3 wells and store the culture medium at 4 °C (*see* Fig. 2).
3. After the final time point collection, Wash 18 wells (6 Tags in triplicates) per time point of a goat anti-mouse IgG coated 96-well microplate 3× with 200 μl wash buffer (*see* **Note 5**).
4. Per timepoint, prepare 150 μl wash buffer with mouse anti-Tag monoclonal IgG (10 μg/ml, *see* Table 1, *see* **Note 6**) for all 6 Tags. Coat the wells with 50 μl anti-Tag monoclonal mixture per well (*see* Fig. 2) and incubate for 1 h at 4 °C spinning at $500 \times g$ followed by 1 h of incubation at room temperature on a microplate shaker at 65–75 RPM (to create a gentle swirl in the wells).
5. Then, aspirate anti-Tag monoclonal mixture and the wash the wells 3× with 200 μl of block buffer for 5 min on a microplate shaker at 65–75 RPM (to create a gentle swirl in the wells).
6. To bind the $Gluc_{Tag}$ from the conditioned culture medium, add 30 μl of the $Gluc_{6x\ Tag}$ conditioned culture medium to a mouse anti-tag coated well and incubate at room temperature

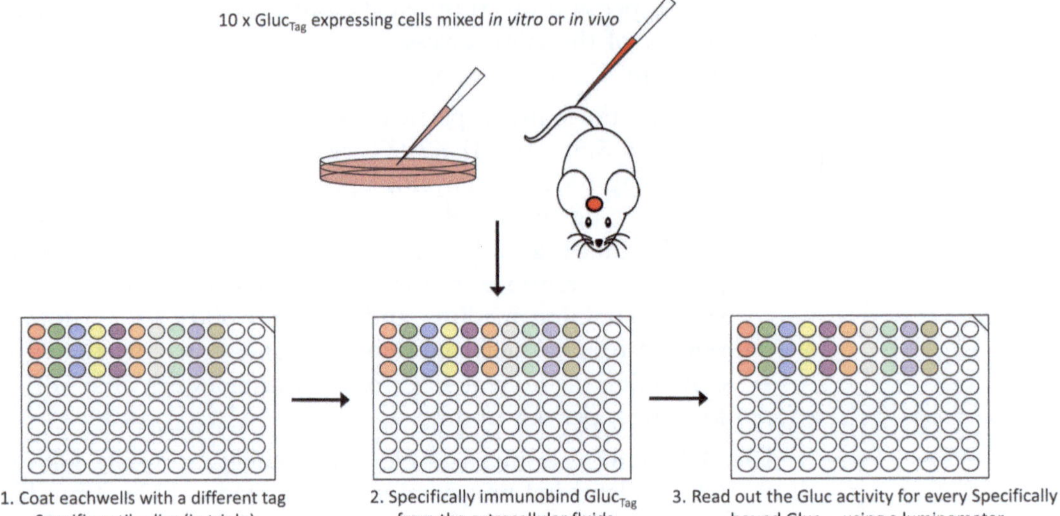

Fig. 3 Immunobinding assay. A white 96-well microplate is coated with anti-tag antibodies. Every well contains one antibody specific for one Gluc$_{Tag}$. Cell culture medium or animal blood containing a mix of all ten different Gluc$_{Tags}$ is added to the well to immunobind a specific Gluc$_{Tag}$ reporter. After washing the bioluminescent activity of every specific Gluc$_{Tag}$ reporter is determined using a luminometer

for 2 h on a microplate shaker with 65–75 RPM (to create a gentle swirl in the wells).

7. After this, aspirate the culture medium and wash the wells 5× for 5 min with 200 μl wash buffer on a microplate shaker at 65–75 RPM (creating a gentle swirl in the wells).
8. To measure the bound Gluc$_{Tag}$ in the mouse anti-Tag coated wells, aspirate the wash buffer completely and just before measurement add 50 μl coelenterazine Gluc substrate per well using a multichannel pipette. Immediately insert the plate in the microplate luminometer (*see* **Note 11**), shake the plate briefly and read out the wells for 0.1 s per well (*see* Fig. 3).

3.2.2 Gluc$_{Tag}$ Immunostaining In Vitro

1. Plate 10^5 of the U87-Gluc$_{tag}$ cell lines (*see* **Note 10**) in a 24 well plate overnight in 500 μl DMEM complete culture medium at 37 °C and 5 % CO_2.
2. The next day, wash the cells with 200 μl PBS and fix with 100 μl 3.7 % formaldehyde fixative solution for 20 min.
3. Then, wash and block the cells 3×5 min with 200 μl block buffer and permeabilize the cells with 100 μl permeabilization buffer for 15 min.
4. Next, wash and block the cells 2×5 min with 200 μl block buffer.
5. Add 100 μl primary mouse anti-tag IgG (2 μg/ml, *see* **Note 6**) in block buffer to the cells and incubate for 1 h at room temperature.

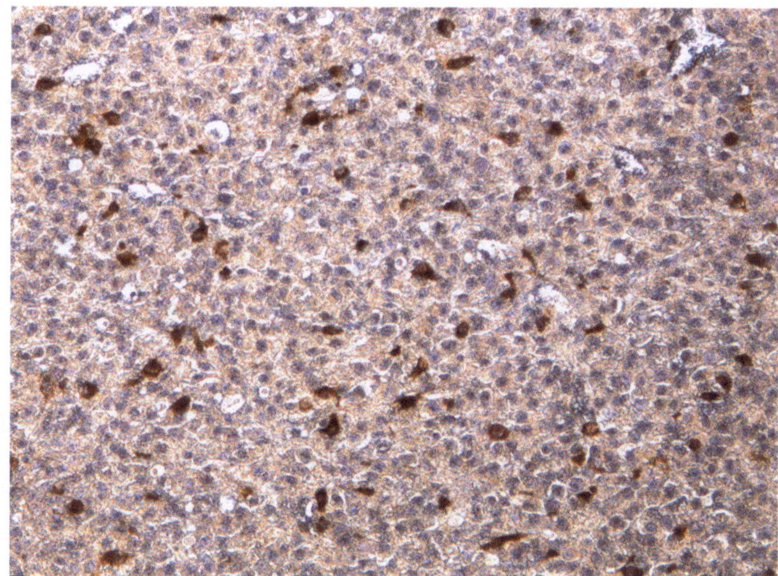

Fig. 4 $Gluc_{Tag}$ immunostaining. Representative immunostaining of ten mixed U87-$Gluc_{Tag}$-CFP cell lines in a mouse xenograft, each expressing another $Gluc_{Tag}$. In this example, an immunostaining against $Gluc_{V5}$ was performed using an anti-V5 mouse immunoglobulin. The $Gluc_{V5}$ expressing U87 cells can be clearly distinguished from the other cells in the tumor

6. Subsequently, wash and block the cells 3×5 min with 200 μl block buffer.
7. Add 100 μl secondary goat anti-mouse-HRP antibody (5 μg/ml, *see* **Note 12**) in block buffer to the cells and incubate for 1 h at room temperature.
8. Then, wash the cells 3× for 5 min with 200 μl block buffer and stain the cells with the DAB reagent set according to the manufacturer's guidelines. Analyze the immunostaining using light microscopy (*see* Fig. 4).

3.2.3 $Gluc_{Tag}$ Immunobinding Assay In Vivo

1. Culture and collect the U87-$Gluc_{Tag}$-CFP cells as described before. Collect 5×10^5 U87-$Gluc_{Tag}$-CFP cells per orthotopical injection in athymic nude-foxn1^{nu} mouse. For experimental triplicates, inject three mice.
2. 30 min before surgery of the mouse, subcutaneously administer Temgesic (Buprenorphine hydrochloride, 0.1 mg/kg) in PBS analgesia and anesthetize the mouse using an isoflurane anesthesia and fix the mouse in a small animal stereotaxic frame.
3. Incise the head skin with a surgical scalpel and locate the injection site $x=2$ mm and $y=0.5$ mm from bregma. Drill a hole in the skull and slowly inject the U87-$Gluc_{Tag}$-CFP cell

suspension into the brain. Then, cover up the skull with the head skin and stitch the wound.

4. At predetermined time points, collect 100 μl of the $Gluc_{Tag}$ conditioned mouse blood of all the mice using a capillary collection and sample container containing EDTA to prevent blood clotting and store the blood at 4 °C (*see* **Note 7**).
5. After the final blood sample collection, dilute the blood 1:1 with blood dilution buffer to increase the sample volume to 180 μl.
6. Wash 18 wells (6 Tags in triplicates) per time point of a goat anti-mouse IgG coated 96-well microplate (*see* **Note 5**) 3× with 200 μl wash buffer.
7. Per timepoint, prepare 150 μl wash buffer with mouse a nti-Tag monoclonal IgG (10 μg/ml, *see* Table 1) for all 6 Tags (*see* **Note 13**). Coat the wells with 50 μl anti-Tag monoclonal mixture per well and incubate for 1 h at 4 °C spinning at 500 × *g* followed by 1 h of incubation at room temperature.
8. Then, aspirate anti-Tag monoclonal mixture and the wash the wells 3× with 200 μl of block buffer for 5 min on a microplate shaker at 65–75 RPM (to create a gentle swirl in the well).
9. To bind the $Gluc_{Tag}$ from the conditioned mouse blood, add 30 μl of the diluted $Gluc_{6x\ Tag}$ blood samples to a mouse anti-tag coated well and incubate at room temperature for 2 h on a microplate shaker with 65–75 RPM (to create a gentle swirl in the well).
10. After this, aspirate the diluted $Gluc_{6x\ Tag}$ blood samples and wash the wells 5× for 5 min with 200 μl wash buffer on a microplate shaker at 65–75 RPM (creating a gentle swirl).
11. To measure the bound $Gluc_{Tag}$ in the mouse anti-Tag coated wells, aspirate the wash buffer completely and just before measurement add 50 μl coelenterazine Gluc substrate per well using a multichannel pipette. Immediately insert the plate in the microplate luminometer plate reader, shake the plate briefly and read out the wells for 0.1 s per well (*see* Fig. 3 and **Note 11**).

3.2.4 $Gluc_{Tag}$ Immunostaining In Vivo

1. After the in vivo $Gluc_{Tag}$ assay, collect and fix the mouse brains containing the tumor in 3.7 % formaldehyde for at least 48 h. Then, dehydrate and embed the tissue samples in paraffin.
2. Using a microtome, section the paraffin embedded tissue samples in 5 μm slices and mount the slices on glass slides.
3. Deparaffinize the tissue slices in xylene and rehydrate in a series of 100 % ethanol, 96 % ethanol, and 75 % ethanol.
4. Block the endogenous peroxidase with fresh 0.3 % H_2O_2 in methanol for 30 min at room temperature and rinse the slices with H_2O.

5. Perform antigen retrieval by cooking the slices in citrate buffer (pH 6) using a microwave. Cook the samples 5 min at 100 % power and 10 min at 50 % power. Cool the slices down to room temperature in 20 min and rinse the slices 3 × 5 min with PBS.
6. Incubate the slices with mouse anti-Tag monoclonal IgG (final concentration is 2 μg/ml, *see* **Note 6**) diluted in antibody diluent for 1 h at room temperature. After IgG incubation, rinse the slices 3 × 5 min with PBS.
7. Incubate the slices with anti-mouse-HRP (1 in 200) diluted in antibody diluent for 30 min at room temperature and rinse the slices 3 × 5 min in PBS.
8. Incubate the slices with DAB for 5–10 min and rinse the slices with H_2O.
9. Counterstain the cell nuclei of the tissue slices with hematoxylin for 30–60 s and rinse the slices with PBS.
10. Dehydrate the slices in the ethanol series of 75 % ethanol, 96 % ethanol and 100 % ethanol and incubate the slices in xylene for 5 min. Let the tissue slices dry and analyze them using a light microscope (*see* Fig. 4).

4 Notes

1. In the assay described here, we use the CSCW-Gluc-IRES-CFP lentiviral vector DNA as a template to amplify the Gluc reporter gene and to religate our Gluc_{Tag} reporter constructs in. When using other template vectors or reporters, it is necessary do redesign primers, restriction sites and oligonucleotide sequences and PCR programs should be optimized in order to fit the alternative vector DNA or reporter gene.
2. We advise to use a high fidelity proof reading DNA polymerase for the amplification of the Gluc reporter gene. This minimizes optimization and the risk of copy errors due to incorrect base-pairing.
3. We advise to use high fidelity restriction enzymes for restriction of the DNA vectors. This increases restriction efficiency and user simplicity since most high fidelity enzymes are optimized in the same restriction reaction buffer. Also, most high fidelity restriction enzymes are optimized to have less star activity, increasing ligation efficiency in later steps of the protocol.
4. There are less hazardous alternatives available for ethidium bromide, such as SYBR safe DNA gel stain (Life Technologies). Besides being less hazardous, these stains can also be used to visualize DNA with non-UV-light, decreasing damage of the DNA due to UV exposure.

5. For luciferase assays it is best to use white microplates to prevent crossover detection of photons between wells. The white goat anti-mouse IgG coated microplates (Thermo Scientific) are pre-coated to improve assay stability but it is also possible to manually coat microplates (e.g., ELISA microplates).
6. The antibodies we used are all mouse monoclonal IgG antibodies. The coated microplates used (*see* **Note 5**) are optimized to use with mouse IgG antibodies. Using monoclonal antibodies over polyclonal antibodies improves assay stability. We were able to optimize 6 gluc_{Tag} antibody combinations for high performance binding capacity while for Gluc_{Tag} antibody combinations showed low binding capacity due to a lower affinity of the antibody to the epitope tag. When using new antibodies, you should optimize the immunobinding assay by validating the antibody binding capacity to the Gluc_{Tag}.
7. The capillary collection and sample container (Sarstedt) we used combines a capillary blood collector with an EDTA coated sample container. This enables fast and easy sample collection and storage. If you collect mouse blood using other methods, make sure you add EDTA to the blood sample to prevent clotting.
8. To quantify the concentration of the DNA gel extracts, we do not recommend the use of spectrophotometry (NanoDrop) since the concentration will usually be low (<100 ng/μl). We advise to quantify the concentrations of the DNA gel extracts by loading a small amount of DNA sample on a gel and quantify the resulting bands by comparing to the molecular marker. We used ImageJ to quantify the DNA sample concentrations and calculate the ligation conditions.
9. Generally, higher virus titers are obtained using other methods of transfection of the plasmids and vectors. Using Lipofectamine 2000 (Life Technologies) or Fugene (Promega) can be beneficial for virus production titers.
10. It is possible to transduce other cell lines of interest but not all cell lines are equally resistant to lentiviral transduction and therefore the transduction conditions need to be optimized for every cell line. To increase transduction efficiency, it might be beneficial to culture the cells with polybrene (2–10 μg/ml) before transduction. Also using polybrene requires optimization, depending on the cell line of use. Expression of the reporters is also cell line dependent since the CMV promoter is not equally active in all cell lines and therefore also the promoter might be an issue for optimization.
11. Since the Gluc photon signal is degrading over time (~10 % per min), the most stable method of measuring would be to measure Gluc signal directly after coelenterazine substrate addition.

A plate reader with a substrate injector would be optimal but it is also an option to use a multichannel pipette. Measuring a 96-well microplate for 0.1 s per well would take about 10 s, so the time difference as a result of measuring the first well and the last well would be well within 10 % deviation as a result of the Gluc signal degradation.

12. For this assay we used goat anti-mouse secondary Ig. It would also be possible to use other anti-mouse Ig antibodies but optimization is required.
13. As a negative control for the immunobinding assays and the stainings it is possible to use the CSCW-Gluc$_{ctrl}$-CFP vector. This vector contains the Gaussia luciferase not fused to an epitope tag.

References

1. Badr CE, Tannous BA (2011) Bioluminescence imaging: progress and applications. Trends Biotechnol 3:1–10. doi:10.1016/j.tibtech.2011.06.010
2. Stratowa C, Audette M (1995) Transcriptional regulation of the human intercellular adhesion molecule-1 gene: a short overview. Immunobiology 193:293–304. doi:10.1016/S0171-2985(11)80558-9
3. Wrana JL, Attisano L, Wieser R et al (1994) Mechanism of activation of the TGF-beta receptor. Nature 370:341–347. doi:10.1038/370341a0
4. Subramaniam D, Natarajan G, Ramalingam S et al (2008) Translation inhibition during cell cycle arrest and apoptosis: Mcl-1 is a novel target for RNA binding protein CUGBP2. Am J Physiol Gastrointest Liver Physiol 294:G1025–G1032. doi:10.1152/ajpgi.00602.2007
5. Niers JM, Kerami M, Pike L et al (2011) Multimodal in vivo imaging and blood monitoring of intrinsic and extrinsic apoptosis. Mol Ther 19:1090–1096
6. De Wet JR, Wood KV, DeLuca M et al (1987) Firefly luciferase gene: structure and expression in mammalian cells. Mol Cell Biol 7:725–737
7. Loening AM, Wu AM, Gambhir SS (2007) Red-shifted Renilla reniformis luciferase variants for imaging in living subjects. Nat Methods 4:641–643. doi:10.1038/nmeth1070
8. Tannous BA, Kim D-E, Fernandez JL et al (2005) Codon-optimized Gaussia luciferase cDNA for mammalian gene expression in culture and in vivo. Mol Ther 11:435–443. doi:10.1016/j.ymthe.2004.10.016
9. McNabb DS, Reed R, Marciniak RA (2005) Dual luciferase assay system for rapid assessment of gene expression in Saccharomyces cerevisiae. Eukaryot Cell 4:1539–1549. doi:10.1128/EC.4.9.1539-1549.2005
10. Van Rijn S, Nilsson J, Noske DP et al (2013) Functional multiplex reporter assay using tagged Gaussia luciferase. Sci Rep 3:1046. doi:10.1038/srep01046
11. Ford CF, Suominen I, Glatz CE (1991) Fusion tails for the recovery and purification of recombinant proteins. Protein Expr Purif 2:95–107
12. Dull T, Zufferey R, Kelly M et al (1998) A third-generation lentivirus vector with a conditional packaging system. J Virol 72:8463–8471

Chapter 19

Noninvasive In Vivo Monitoring of Extracellular Vesicles

Charles P. Lai, Bakhos A. Tannous, and Xandra O. Breakefield

Abstract

Extracellular vesicles (EVs) including exosomes and microvesicles are nanometer-sized vesicles released by cells to deliver lipids, cellular proteins, mRNAs, and noncoding RNAs, thereby facilitating intercellular communication without direct cell-to-cell contacts. Due to their nanoscale size, EVs have been visualized under microscopy in vitro. We here describe a strategy to label EVs with *Gaussia* luciferase for noninvasive bioluminescence imaging and monitoring of systemically administered EVs in vivo.

Key words Extracellular vesicles, Microvesicles, Exosomes, *Gaussia* luciferase, Bioluminescence, Coelenterazine

1 Introduction

Cells release various types of extracelluar vesicles (EVs) ranging from 40 to 4,000 nm in diameter, which are capable of transporting lipids, DNAs, RNAs, and proteins to recipient cells. These vesicles include exosomes (40–100 nm), microvesicles (100–1,000 nm), and apoptotic bodies (50–4,000 nm) [1, 2]. Since EVs are diffuse within the extracellular matrix and taken up by cells, they can mediate intercellular communication between neighboring cells, as well as cells at distal sites [3–9]. Importantly, cancer cells were found to produce an abundant amount of EVs containing a select subset of cellular proteins, mRNAs and noncoding RNAs, thereby modulating normal cells in their microenvironment to promote angiogenesis and tumor progression [10–20]. On the other hand, since EVs are small and poorly immunogenic, several recent studies have engineered EVs to shield therapeutic genes/drugs for targeted delivery to tumors, even overcoming biological barriers such as the blood–brain barrier [21–26]. Until recently, EVs have only been visualized under high-resolution microscopy due to their nanometer size, and localizing EVs in vivo has been largely unachievable [27].

Christian E. Badr (ed.), *Bioluminescent Imaging: Methods and Protocols*, Methods in Molecular Biology, vol. 1098, DOI 10.1007/978-1-62703-718-1_19,

Photinus pyralis luciferase (firefly; Fluc), *Renilla reniformis* luciferase (sea pansy; Rluc), and *Gaussia princeps* luciferase (marine copepod; Gluc) are among the most commonly used luciferases in biomedical research [28]. Unlike reporters conjugated to fluorescent proteins, such as CD63-GFP for EVs [19, 29], luciferase-based reporters do not need an excitation source for their subsequent light emission. Instead, luciferases emit bioluminescence via conversion of their respective substrate with either ATP and Mg^{2+} (Fluc; *D*-luciferin), or oxygen alone (Rluc and Gluc; coelenterazine). This makes luciferase an ideal reporter for in vivo studies where specialized equipment for excitation of reporter proteins is not required.

Gluc is unique from other luciferases such that it is naturally secreted and capable of emitting flash bioluminescence (480 nm peak) that is over 1,000-fold more sensitive than Rluc and Fluc [30]. Combining these features and the fact that Gluc only requires oxygen as its cofactor for bioluminescence reaction, we engineered a membrane-bound (mb) variant of Gluc fused to a biotin acceptor peptide (BAP), termed mbGluc-BAP [31], for EV labeling. By overexpressing mbGluc-BAP in human embryonic kidney (HEK) 293T cells followed by EV isolation, one can successfully track systemically administered EVs in athymic nude mice via in vivo bioluminescence imaging or by ex vivo analysis of tissues.

2 Materials

2.1 EV Reporter Construct and Donor Cells

1. EV reporter construct: pCSCW2-mbGluc-BAP-IRES-GFP, a lentivirus vector encoding Gluc fused to the transmembrane domain of platelet-derived growth factor receptor (PDGFR) and a biotin acceptor protein (mbGluc-BAP) as well as GFP, separated by an internal ribosomal entry site (IRES), all under the control of the constitutively active cytomegalovirus (CMV) promoter [32].
2. EV donor cells: Human embryonic kidney 293T (293T) cells obtained from American Type Culture Collection (ATCC; Manassas, VA, USA). These cells are cultured in Dulbecco's modified Eagle's medium (DMEM) supplemented with 10 % fetal bovine serum (FBS), 100 U/ml penicillin, and 0.1 mg/ml streptomycin at 37 °C and 5 % CO_2.
3. Filter: 0.22 μm polyethersulfone filter (EMD Millipore, Billerica, MA, USA).
4. EV-depleted FBS: Using ultracentrifugation equipment described (*see* Subheading 2.2, **items 1–3**), centrifuge FBS at 100,000 × *g* for at least 16 h at 4 °C. Collect supernatant (EV-depleted FBS) and filter it through a 0.22 μm filter to

further minimize EV contamination of the FBS. Aliquot EV-depleted FBS into 50 ml canonical tubes and store at −80 °C.

5. EV-isolation medium: Supplement DMEM with prepared 10 % EV-depleted FBS, 100 U/ml penicillin and 0.1 mg/ml streptomycin.
6. Polybrene: 10 mg/ml in tissue culture grade water.
7. Epifluorescence microscope capable of detecting GFP expression.

2.2 EV Isolation Components

1. Ultracentrifuge: Optima L-90K (Beckman-Coulter, Indianapolis, IN, USA).
2. Rotor: Type 70Ti rotor (Beckman-Coulter).
3. Ultracentrifuge tubes, seal former, tube sealer: Quick-Seal® tubes (Cat#: 342413, Beckman-Coulter), seal former and a tube sealer (Beckman-Coulter).
4. Filter: 0.22 μm polyethersulfone filter (EMD Millipore).
5. 60 ml syringes.
6. Double filtered phosphate buffered saline (PBS): 8 g/L NaCl, 0.2 g/L KCl, 1.78 g/L $Na_2HPO_4 \cdot 2\ H_2O$, 0.27 g/L KH_2PO_4, adjust to pH 7.4 and filter twice through a 0.22 μm filter.

2.3 Gluc Activity Assay

1. A luminometer such as the Dynex MLX Microtiter Plate Luminometer (Vienna, VA, USA).
2. 96-well white or black microtiter plates.
3. Coelenterazine (CTZ), the Gluc substrate: Reconstitute CTZ to a stock concentration of 5 mg/ml with methanol pre-supplemented with a few drops of HCl to minimize CTZ oxidation and precipitation. Aliquot CTZ and store at −20 °C. Pre-warm CTZ to room temperature before use and protect from light.

2.4 In Vivo Bioluminescence Imaging

1. Animals: Immunodeficient athymic nude mice or any other mouse strain of choice.
2. An optical imaging system such as the IVIS® Spectrum connected to XGI-8 Anesthesia System (Perkin-Elmer, Waltham, MA, USA).
3. Isoflurane (Baxter, Deerfield, IL, USA).
4. XGI-8 Anesthesia System (Perkin-Elmer).
5. Sterile PBS: Autoclaved or 0.22 μm filtered PBS.
6. Coelenterazine: prepare as described in **item 3** in Subheading 2.3. Alternatively, for more stable Gluc signal during imaging, prepare fresh CTZ before each injection according to “Inject-A-Lume” kit manual (Nanolight, Pinetop, AZ, USA).
7. 1.5 ml microcentrifuge tubes.

8. Insulin syringe needles: 29½ gauge, 0.3 ml.
9. Heating device: 43 °C water in a 500 ml beaker to induce vasodilation of tail veins.

3 Methods

3.1 Generate EV Donor Cells

1. Seed 0.5×10^6 HEK 293T cells per well in fresh culturing medium in a 6-well tissue culture plate (34.8 mm diameter).
2. When 60–70 % confluence is reached, transduce the cells with lentiviral vectors encoding mbGluc-BAP-IRES-GFP in 3 ml medium supplemented with 1 μl polybrene (10 mg/ml) for each well (*see* **Note 1**).
3. 48 h post-transduction, examine GFP reporter expression under an epifluorescence microscope to determine transduction efficiency. At least 80–90 % of cells should be transduced to generate sufficient EVs labeled with mbGluc-BAP (*see* **Note 2**).

3.2 EV Isolation and Gluc Activity Assay

1. Add 2×10^6 of 293T cells expressing mbGluc-BAP per 150 mm plate in fresh medium.
2. When the cells reach approximately 50 % confluence, gently aspirate the growth medium and replace with 20 ml of fresh EV-isolation medium.
3. Incubate the cells for another 48 h at 37 °C, 5 % CO_2 to generate conditioned medium containing EVs.
4. Collect conditioned medium in 50 ml canonical tubes and centrifuge at $300 \times g$ for 10 min at 4 °C to remove cell debris.
5. Collect supernatant in separate 50 ml canonical tubes and centrifuge at $2{,}000 \times g$ for 10 min at 4 °C to remove additional cell debris.
6. Prepare a 60 ml syringe attached to a 0.22 μm filter followed by a 16 gauge needle.
7. Insert the assembled filtering syringe into a Quick-Seal® tube and filter the supernatant through the syringe (*see* **Note 3**).
8. Weigh all supernatant-containing Quick-Seal® tubes and adjust to equal weight by removing excess supernatant as needed and seal the tubes with a sealer. Mark one side of each tube with a marker pen for later orientation purposes (Fig. 1a).
9. In a Type 70 Ti rotor, place tubes with equal weight diagonally opposite to one another for balance. Note orientation of the marked tubes since EV pellets will be collected from the side away from center of the rotor (Fig. 1b). Apply Quick-Seal® spacers and lid for the rotor, centrifuge at $100{,}000 \times g$ for 80 min at 4 °C.

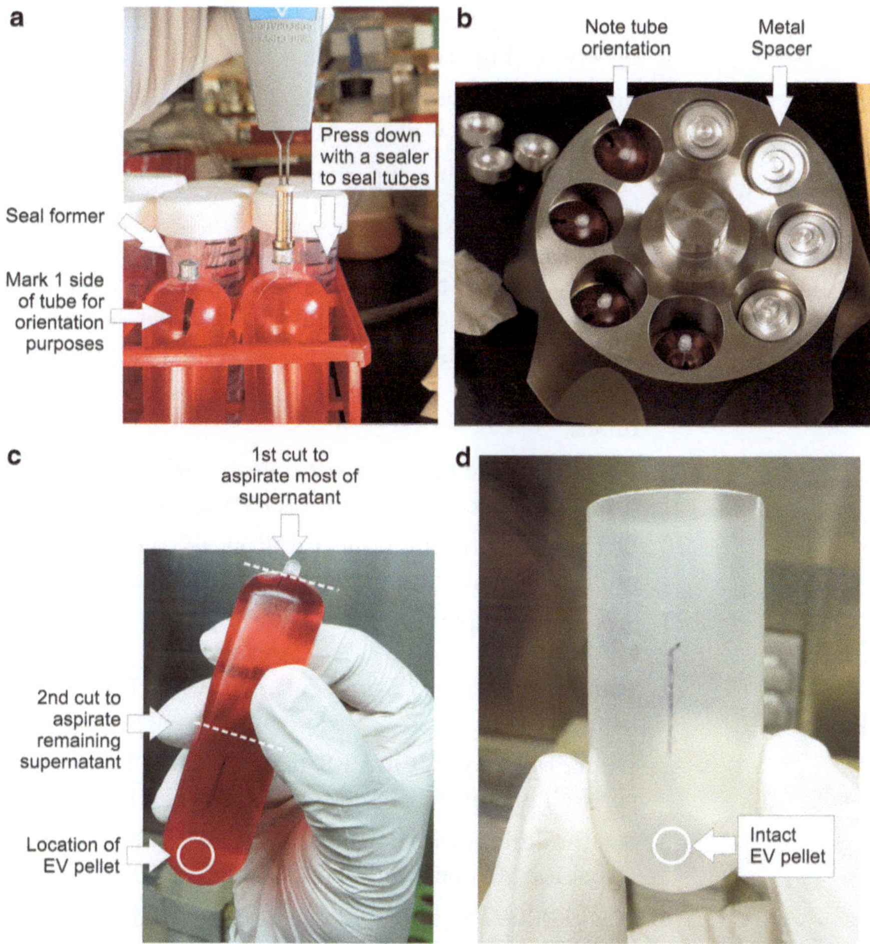

Fig. 1 EV isolation procedure. (**a**) Following conditioned medium collection into ultracentrifuge tubes, place a seal former onto each tube stem. Seal the tubes with a tube sealer by applying gentle, downward pressure until the seal former rests on top of the tube shoulder. Do not press the seal former beyond the tube shoulder since it will result in leakage during centrifugation. Using a permanent marker, mark one side of the tubes (*both top and bottom*) for orientation purposes. (**b**) Place sealed tubes with equal weight opposite to one another in the rotor. Make sure all pre-marked tubes have the same orientation for later EV isolation purposes. Secure the tubes with metal spacers, and firmly seal the rotor with the rotor lid. (**c**) Location of EV pellet and suggested cut sites for staged aspiration to remove supernatant. (**d**) Intact EV pellet after the supernatant is removed

10. Gently remove sealed tubes from the rotor with minimal agitation (*see* **Note 4**).
11. Locate EV pellet at lower side of the tube wall—the side away from the center of the rotor (Fig. 1c).
12. While holding the tube firmly to prevent sudden movement of the tube, carefully cut the tube stem with a razor blade to create an opening (Fig. 1c, 1st cut).
13. Using a Pasteur pipette, gently aspirate most of the supernatant without disturbing the pellet. Focus on the pellet during this procedure to ensure that the pellet is not lost.

14. Cut the tube from the midsection with a razor blade or a pair of scissor (Fig. 1c, 2nd cut).
15. Position the tube upright for 2 min to allow remaining supernatant from the wall to settle to the bottom.
16. Aspirate remaining supernatant with a pipette without disturbing the pellet (*see* **Note 5**). The pellet may or may not be readily visible at this stage (Fig. 1d).
17. Using a 200 μl tip, resuspend EV pellets by pipetting up and down in 100 μl of double filtered PBS per tube.
18. Situate the tubes on ice to keep original pellet spot covered in PBS for 30 min. This step allows better EV recovery from the tubes.
19. Resuspend the pellet by trituration with a pipette and transfer the EV sample into a 1.5 ml microfuge tube.
20. Determine protein concentration of the EV sample using preferred protein quantification methods (*see* **Note 6**).
21. Store the samples at −80 °C.

3.3 Gluc Assay on Isolated EVs

1. Dilute CTZ stock (5 mg/ml) in PBS at 1:1,000 and protect from light.
2. Dilute isolated EVs in PBS at 1:100. Add 10 μl of diluted EVs per well in triplicates into a 96-well white microtiter plate.
3. Set the luminometer to dispense 50 μl CTZ and acquire the signal for 10 s per well, with 2 s integration time.
4. A 100,000–200,000-fold increase in signal should be observed in mbGluc-BAP EVs when compared to negative or blank controls, thereby validating successful EV labeling (*see* **Note 7**).

3.4 In Vivo EV Bioluminescence Imaging

1. For each animal, prepare 100 μg of EVs in a total volume of 200 μl with PBS in a 1.5 ml microfuge tube.
2. Place an animal in a restrainer. Warm the tail with 43 °C water in a beaker for a few minutes until tail veins are readily visible for injection.
3. Inject the EV mixture into the tail vein by using an insulin syringe needle (*see* **Note 8**).
4. For each animal to be imaged, prepare Gluc substrate by adding 20 μl CTZ (5 mg/ml) to 130 μl of PBS in a 1.5 ml microfuge tube, and mix well by pipetting up and down (*see* **Note 9**).
5. 15–30 min following EV injection, anesthetize the animal by isoflurane (1–1.5 % at 2 L/min) using the XGI-8 Anesthesia System.
6. Inject 150 μl CTZ mixture via tail or retro-orbital vein with an insulin syringe needle (*see* **Note 10**).

7. Immediately place the animal in the IVIS® Spectrum connected to the XGI-8 Anesthesia System for continued general anesthesia.
8. Immediately following CTZ administration, image the animal for bioluminescence signal by IVIS® Spectrum (*see* **Note 11**).

4 Notes

1. Determine the amount of lentivirus required for transduction based on multiplicity of infection (MOI). 293T cells are highly prone to lentiviral infection, and a MOI of ≥5 transducing units per cell is recommended. Alternatively, cells can be transfected with plasmids encoding mbGluc-BAP-IRES-GFP for transient expression.
2. Transduced cells can be further sorted for GFP expression by fluorescence-activated cell sorting (FACS) to ensure a high percentage of mbGluc-BAP positive cell population. The transduced/FACS sorted cells can also be frozen for later experiments. Low percentage of positive cells will yield a low amount of mbGluc-BAP-labeled EVs, making them unsuitable for subsequent imaging experiments.
3. Same filtering syringe can be reused between filtrations. Detach syringe from the filter first before removing the plunger between filtrations—this helps in maintaining filter integrity. When the syringe becomes difficult to depress, replace the filter.
4. EV pellets detach easily from tube wall when the tubes undergo sudden movement. Use extra caution when handling the tubes from this step and on.
5. Try to reduce remaining supernatant as much as possible by hand pipetting when approximately 500 μl supernatant is left. Residual supernatant will result in a significant increase in protein concentration (i.e., FBS contamination) and affect Gluc signal in subsequent assays.
6. Dilute the EV samples 1:5, 1:10, and 1:20 in PBS to estimate EV protein concentration. An average EV concentration of 0.5–1 μg/μl is expected.
7. EVs derived from parental 293T cells exhibit no Gluc activity and can therefore be used as a negative control.
8. EVs have a short half-life of <30 min in mice. Therefore timing following EV administration is critical to minimize a loss of EV signal during imaging.
9. When "Inject-A-Lume" kit (Nanolight) is used, apply 150 μl of "Fuel-Inject" diluent to NanoFuel™ vial, and mix well until solution becomes clear.

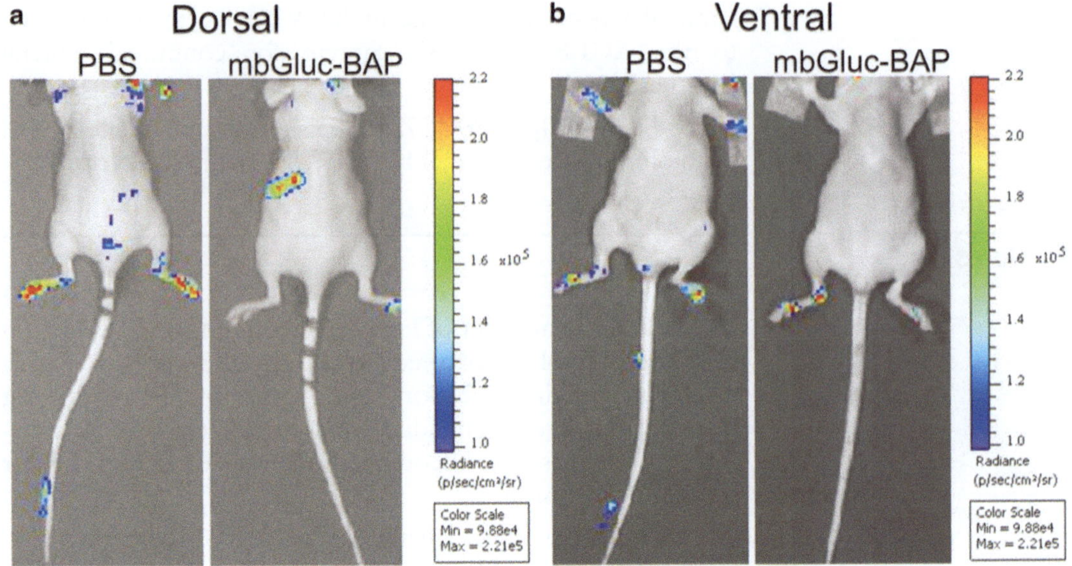

Fig. 2 In vivo bioluminescence imaging of EVs. (**a**) Dorsal side of nude mice administered with either PBS (control) [changed order as read from *left* to *right*] or mbGluc-BAP EVs. The spleen of mbGluc-BAP injected mouse exhibited a prominent bioluminescence signal when compared to the control at 30 min post-EV injection. (**b**) Ventral side of nude mice showing control and mbGluc-BAP EV localizing predominately to the liver and possibly spleen at 30 min post-EV injection. Although not readily visible here, PBS-injected mouse typically exhibited low level bioluminescence from oxidation of CTZ in the liver

10. When using the "Inject-A-Lume" kit, use only 70–75 μl of the CTZ mixture per 25 g mouse. Do not use more than 80 μl of the CTZ mixture from "Inject-A-Lume" kit per 25 g mouse since it is lethal. Timing is absolutely critical following CTZ administration to ensure optimal capture of Gluc signal from mbGluc-BAP labeled EVs. Highest signal can be detected at 1 min post-CTZ injection followed by a rapid loss of signal within 5 min thereafter.
11. Allow 1 min post-CTZ injection to achieve highest Gluc signal intensity. Due to the short half-life of both EVs and CTZ, it is recommended to prepare one animal each for either dorsal (Fig. 2a) or ventral side (Fig. 2b) imaging for optimal imaging signal.

Acknowledgement

We wish to thank Mr. Osama Mardini for his technical assistance. This work was supported by NIH grant NIH/NCI CA069246, CA141150 and Canadian Institutes of Health Research (CIHR).

References

1. Lai CP-K, Breakefield XO (2012) Role of exosomes/microvesicles in the nervous system and use in emerging therapies. Front Physiol 3:228. doi:10.3389/fphys.2012.00228
2. Breakefield XO, Frederickson RM, Simpson RJ (2011) Gesicles: microvesicle "cookies" for transient information transfer between cells. Mol Ther 19:1574–1576. doi:10.1038/mt.2011.169
3. Gross JC, Chaudhary V, Bartscherer K, Boutros M (2012) Active Wnt proteins are secreted on exosomes. Nat Cell Biol. doi:10.1038/ncb2574
4. Fauré J, Lachenal G, Court M et al (2006) Exosomes are released by cultured cortical neurones. Mol Cell Neurosci 31:642–648. doi:10.1016/j.mcn.2005.12.003
5. Kogure T, Lin W-L, Yan IK et al (2011) Intercellular nanovesicle-mediated microRNA transfer: a mechanism of environmental modulation of hepatocellular cancer cell growth. Hepatology 54:1237–1248. doi:10.1002/hep.24504
6. Marzesco A-M, Janich P, Wilsch-Bräuninger M et al (2005) Release of extracellular membrane particles carrying the stem cell marker prominin-1 (CD133) from neural progenitors and other epithelial cells. J Cell Sci 118:2849–2858. doi:10.1242/jcs.02439
7. Mittelbrunn M, Gutiérrez-Vázquez C, Villarroya-Beltri C et al (2011) Unidirectional transfer of microRNA-loaded exosomes from T cells to antigen-presenting cells. Nat Commun 2:282. doi:10.1038/ncomms1285
8. Montecalvo A, Larregina AT, Shufesky WJ et al (2012) Mechanism of transfer of functional microRNAs between mouse dendritic cells via exosomes. Blood 119:756–766. doi:10.1182/blood-2011-02-338004
9. Valadi H, Ekström K, Bossios A et al (2007) Exosome-mediated transfer of mRNAs and microRNAs is a novel mechanism of genetic exchange between cells. Nat Cell Biol 9:654–659. doi:10.1038/ncb1596
10. Skog J, Würdinger T, van Rijn S et al (2008) Glioblastoma microvesicles transport RNA and proteins that promote tumour growth and provide diagnostic biomarkers. Nat Cell Biol 10:1470–1476. doi:10.1038/ncb1800
11. Peinado H, Alečković M, Lavotshkin S et al (2012) Melanoma exosomes educate bone marrow progenitor cells toward a pro-metastatic phenotype through MET. Nat Med. doi:10.1038/nm.2753
12. Luga V, Zhang L, Viloria-Petit AM et al (2012) Exosomes mediate stromal mobilization of autocrine Wnt-PCP signaling in breast cancer cell migration. Cell 151:1542–1556. doi:10.1016/j.cell.2012.11.024
13. Muralidharan-Chari V, Clancy JW, Sedgwick A, D'Souza-Schorey C (2010) Microvesicles: mediators of extracellular communication during cancer progression. J Cell Sci 123:1603–1611. doi:10.1242/jcs.064386
14. Taylor DD, Gercel-Taylor C (2011) Exosomes/microvesicles: mediators of cancer-associated immunosuppressive microenvironments. Semin Immunopathol 33:441–454. doi:10.1007/s00281-010-0234-8
15. Hood JL, San RS, Wickline SA (2011) Exosomes released by melanoma cells prepare sentinel lymph nodes for tumor metastasis. Cancer Res 71:3792–3801. doi:10.1158/0008-5472.CAN-10-4455
16. Antonyak MA, Li B, Boroughs LK et al (2011) Cancer cell-derived microvesicles induce transformation by transferring tissue transglutaminase and fibronectin to recipient cells. Proc Natl Acad Sci USA 108:4852–4857. doi:10.1073/pnas.1017667108
17. Al-Nedawi K, Meehan B, Kerbel RS et al (2009) Endothelial expression of autocrine VEGF upon the uptake of tumor-derived microvesicles containing oncogenic EGFR. Proc Natl Acad Sci USA 106:3794–3799. doi:10.1073/pnas.0804543106
18. Al-Nedawi K, Meehan B, Micallef J et al (2008) Intercellular transfer of the oncogenic receptor EGFRvIII by microvesicles derived from tumour cells. Nat Cell Biol 10:619–624. doi:10.1038/ncb1725
19. Suetsugu A, Honma K, Saji S et al (2012) Imaging exosome transfer from breast cancer cells to stroma at metastatic sites in orthotopic nude mouse models. Adv Drug Deliv Rev :1–8. doi: 10.1016/j.addr.2012.08.007
20. Balaj L, Lessard R, Dai L et al (2011) Tumour microvesicles contain retrotransposon elements and amplified oncogene sequences. Nat Commun 2:180–189. doi:10.1038/ncomms1180
21. Zhang H-G, Kim H, Liu C et al (2007) Curcumin reverses breast tumor exosomes mediated immune suppression of NK cell tumor cytotoxicity. Biochim Biophys Acta 1773:1116–1123. doi:10.1016/j.bbamcr.2007.04.015
22. Sun D, Zhuang X, Xiang X et al (2010) A novel nanoparticle drug delivery system: the anti-inflammatory activity of curcumin is enhanced when encapsulated in exosomes. Mol Ther 18:1606–1614. doi:10.1038/mt.2010.105
23. Alvarez-Erviti L, Seow Y, Yin H et al (2011) Delivery of siRNA to the mouse brain by systemic injection of targeted exosomes. Nat Biotechnol :1–7. doi: 10.1038/nbt.1807
24. Zhuang X, Xiang X, Grizzle W et al (2011) Treatment of brain inflammatory diseases by delivering exosome encapsulated

anti-inflammatory drugs from the nasal region to the brain. Mol Ther 19:1769–1779. doi:10.1038/mt.2011.164

25. Mizrak A, Bolukbasi MF, Ozdener GB et al (2012) Genetically engineered microvesicles carrying suicide mRNA/protein inhibit schwannoma tumor growth. Mol Ther. doi:10.1038/mt.2012.161
26. Bolukbasi MF, Mizrak A, Ozdener GB et al (2012) miR-1289 and "Zipcode-" like sequence enrich mRNAs in microvesicles. Mol Ther Nucleic Acids 1:e10. doi:10.1038/mtna.2011.2
27. Ohno S-I, Takanashi M, Sudo K et al (2012) Systemically injected exosomes targeted to EGFR deliver antitumor MicroRNA to breast cancer cells. Mol Ther :1–7. doi: 10.1038/mt.2012.180
28. Badr CE, Tannous BA (2011) Bioluminescence imaging: progress and applications. Trends Biotechnol 29:624–633. doi:10.1016/j.tibtech.2011.06.010
29. Koumangoye RB, Sakwe AM, Goodwin JS et al (2011) Detachment of breast tumor cells induces rapid secretion of exosomes which subsequently mediate cellular adhesion and spreading. PLoS One 6:e24234. doi:10.1371/journal.pone.0024234.g010
30. Tannous B, KIM D, Fernandez J et al (2005) Codon-optimized luciferase cDNA for mammalian gene expression in culture and in vivo. Mol Ther 11:435–443. doi:10.1016/j.ymthe. 2004.10.016
31. Tannous BA, Grimm J, Perry KF et al (2006) Metabolic biotinylation of cell surface receptors for in vivo imaging. Nat Methods 3:391–396. doi:10.1038/nmeth875
32. Niers JM, Chen JW, Lewandrowski G et al (2012) Single reporter for targeted multimodal in vivo imaging. J Am Chem Soc 134:5149–5156. doi:10.1021/ja209868g

Chapter 20

In Vitro and In Vivo Demonstrations of Fluorescence by Unbound Excitation from Luminescence (FUEL)

Joe Dragavon, Abdessalem Rekiki, Ioanna Theodorou, Chelsea Samson, Samantha Blazquez, Kelly L. Rogers, Régis Tournebize, and Spencer Shorte

Abstract

Bioluminescence imaging is a powerful technique that allows for deep-tissue analysis in living, intact organisms. However, in vivo optical imaging is compounded by difficulties due to light scattering and absorption. While light scattering is relatively difficult to overcome and compensate, light absorption by biological tissue is strongly dependent upon wavelength. For example, light absorption by mammalian tissue is highest in the blue-yellow part of the visible energy spectrum. Many natural bioluminescent molecules emit photonic energy in this range, thus in vivo optical detection of these molecules is primarily limited by absorption. This has driven efforts for probe development aimed to enhance photonic emission of red light that is absorbed much less by mammalian tissue using either direct genetic manipulation, and/or resonance energy transfer methods. Here we describe a recently identified alternative approach termed Fluorescence by Unbound Excitation from Luminescence (FUEL), where bioluminescent molecules are able to induce a fluorescent response from fluorescent nanoparticles through an epifluorescence mechanism, thereby significantly increasing both the total number of detectable photons as well as the number of red photons produced.

Key words FUEL, BRET, Quantum dot, Bioluminescence, In vivo

1 Introduction

Whole animal bioluminescence imaging (BLI) provides a powerful technological means to observe physiological events inside living systems with minimal perturbation [1–5]. It therefore has utility in studies on, for example, tumor development [3, 6], immunity [4], infection [7], gene expression [8], and inflammation [9]. However, a critical challenge for in vivo BLI comes from an abundance of endogenous optical absorbers, such as hemoglobin [2, 10, 11]. Hemoglobin readily absorbs blue-yellow light (400–650 nm); strongly overlapping the most commonly used bioluminescent

Christian E. Badr (ed.), *Bioluminescent Imaging: Methods and Protocols*, Methods in Molecular Biology, vol. 1098, DOI 10.1007/978-1-62703-718-1_20, © Springer Science+Business Media New York 2014

sources (i.e., *Photorhabdus luminescens*, *Gaussia*, and even firefly luciferase to some extent). Consequently, the light output of luminescent probes can be quenched, and/or scattered by tissues, eventually resulting in a reduced amount of detectable signal. Likewise, these same constraints are encountered under in vitro conditions using animal extracts such as blood or homogenized tissue. Because biological tissue is more permissive to red light, much effort has been directed towards improving BLI sensitivity by red-shifting the emission wavelengths to the desired optical window of 650–900 nm, a range in which optical absorption is minimized significantly [12]. To achieve this, researchers have tried either to create mutations of the common luciferases (for example, the work by Branchini et al.) [12], or by proxy using bioluminescence resonance energy transfer (BRET) [13–15].

Unlike classical fluorescence excitation, where excitation light radiatively excites a fluorophore, BRET is a process by which the photon energy of a luminescent "donor" molecule can pass directly (via a non-radiative resonance energy transfer, RET) to a fluorescent "acceptor" molecule within 10 nm proximity, resulting in fluorescence. The Stokes shift of the fluorophore yields fluorescence emission at a red-shifted, longer wavelength than the original luminescence. Recently we reported an alternative method to achieve a similar red shifted emission by using Fluorescence by Unbound Excitation from Luminescence (FUEL) [2, 10]. The FUEL method differs from BRET because it arises from an irradiative epifluorescent excitation/emission interaction between the bioluminescent molecule and fluorophore. The experimental conditions favoring FUEL are very similar to those for BRET: a strong spectral overlap between the bioluminescent source and the fluorophore must exist. The major difference is the distance tolerance over which the phenomena may occur. The working distance of FUEL is much greater than 10 nm. Under in vitro conditions and in the absence of an optical absorber, the FUEL effect has been observed with up to 3 cm between the luminescent source and the fluorophore moieties, without covalent coupling chemistries or surface modifications of either of the FUEL components.

The differences between FUEL and BRET underline their distinct utility as experimental tools. While the use of BRET is well established in the literature, we propose that FUEL brings a different utility to the experimental toolbox for both in vitro and in vivo applications. For example, the possibility to measure entity co-proximity at a macroscopic level would be invaluable. Towards these ends, we present in detail simple and robust methods to reconstitute FUEL using bioluminescent *Escherichia coli* and quantum dots, under controlled in vitro and in vivo demonstrations. With the principles demonstrated here it should be possible to extrapolate the appropriate techniques for more complicated in vitro or in vivo applications thereof.

2 Materials

1. Standard bacteria culture media (LB for example).
2. Bioluminescent *E. coli* carrying the plasmid expressing the *lux* operon (DH5alpha carrying pUC18-mini-Tn7T-Gm-lux; referred heron as RT57) with an $OD_{600} = 2$ (for the RT57 *E. coli*, an $OD_{600} = 1$ is 5×10^8 bacteria per ml, *see* **Note 1**). Other luminescent bacterial sources may be used such as the commercially available *Staphylococcus aureus* Xen 36 (*PerkinElmer*, USA).
3. Physiological saline (0.09 % NaCl, PS); may be prepared in-house or purchased.
4. Qtracker 705 (2 μM, QD705) non-targeted quantum dots (*Invitrogen*). Keep in the dark at 4 °C until use. The QD705 are used as received.
5. A black-walled 96-well plate.
6. Prepare a positive displacement pipette (*see* **Note 2**).
7. Agarose (*Sigma*), prepared into 3 % and 4 % solutions.
8. For all animal experiments, female 6-week-old BALB/c mice (*Janvier*, France) are used.
9. An electric razor equipped with a fine blade.
10. Commercial epilation cream such as Nair®.
11. Standard chemical anesthesia solution, used at 1.5 mg/kg Ketamine and 30 mg/kg Xylazine.
12. All animal experiments should be carried out only under strict accordance with applicable national and institutional guidelines and only by suitably accredited, qualified personnel.

2.1 Bioluminescence Imaging

1. An IVIS 100 whole animal imaging system (*Xenogen Corporation, Caliper Life Sciences, a PerkinElmer Company*, Alameda, CA) equipped with an intensified deep-cooled CCD and a filter wheel is used to acquire all bioluminescent and fluorescent images.
2. One location within the filter wheel is kept empty for total light detection (Total Light).
3. A 480 nm ± 10 nm band pass filter is used to collect the RT57-specific photons.
4. A Cy5.5 band pass filter (695–770 nm) is used to distinguish red photons.
5. The CCD is cooled to −105 °C and the stage warmed to 37 °C for each experiment.
6. For each image, the acquisition settings for the Living Image (*Xenogen*) software version 3.1 were set as follows: 8× bin, field of view B (20 cm), and f-stop 1.

7. The acquisition time is typically 10–60 s, but can be adjusted as necessary.
8. The illumination power was set to low, the bin setting was reduced to 4×, and the acquisition time set to 0.5 s for fluorescence image acquisitions.
9. QD705 fluorescence was observed using a Cy5.5 filter set that included a 615–665 nm excitation filter and a 695–770 nm emission filter (*see* **Note 3**).

3 Methods

3.1 Demonstration of FUEL in Suspension

This initial experiment attempts to demonstrate the basic building blocks of the FUEL phenomenon. The existence of resonance energy transfer between two unbound moieties is minimal. The event becomes even more unlikely when bacteria are the luminescent source. This is due to the fact that the luminescence evolves from below the inner membrane of the bacteria and that the fluorophore, QD705 in this case, remains exterior to the bacteria obscuring any possibility of binding between the two components occurs. Given the need for the luminescent/fluorescent molecules to be within 10 nm proximity for RET to occur, one can conclude that any detected red signal is due exclusively to FUEL.

1. Prepare an overnight culture of the appropriate bacteria at the appropriate temperature, typically reaching an OD = 2–3 by the next morning. Then, seed subcultures as needed in fresh media until OD ~ 1/mid log phase is reached. Finally, dilute or concentrate if necessary to reach a final concentration of 5×10^8 bacteria/ ml
2. Label three tubes: "Bac," "QD705," "Bac + QD705"
3. Add 90 μl of PBS or physiological saline (PS) and 10 μl of the bacterial stock to the Bac tube. Then, into the QD705 tube, add 95 μl of PS and 5 μl of QD705. Finally, add 85 μl of PS, 10 μl of the bacteria stock solution, and 5 μl of QD705 to the Bac + QD705 tube.
4. Distribute the prepared tubes into individual wells of a black-walled 96-well plate (Fig. 1).
5. Repeat **steps 2** and **3** for as many replicates as desired using a unique bacterial culture each time.
6. Place the black-walled 96-well plate into the IVIS100 and acquire images under the Total Light, 470–490 nm band pass, and the 695–770 nm band pass filters.
7. Confirm QD location using epifluorescence using a 615–665 nm excitation filter and a 695–770 nm emission filter (*see* Fig. 1).

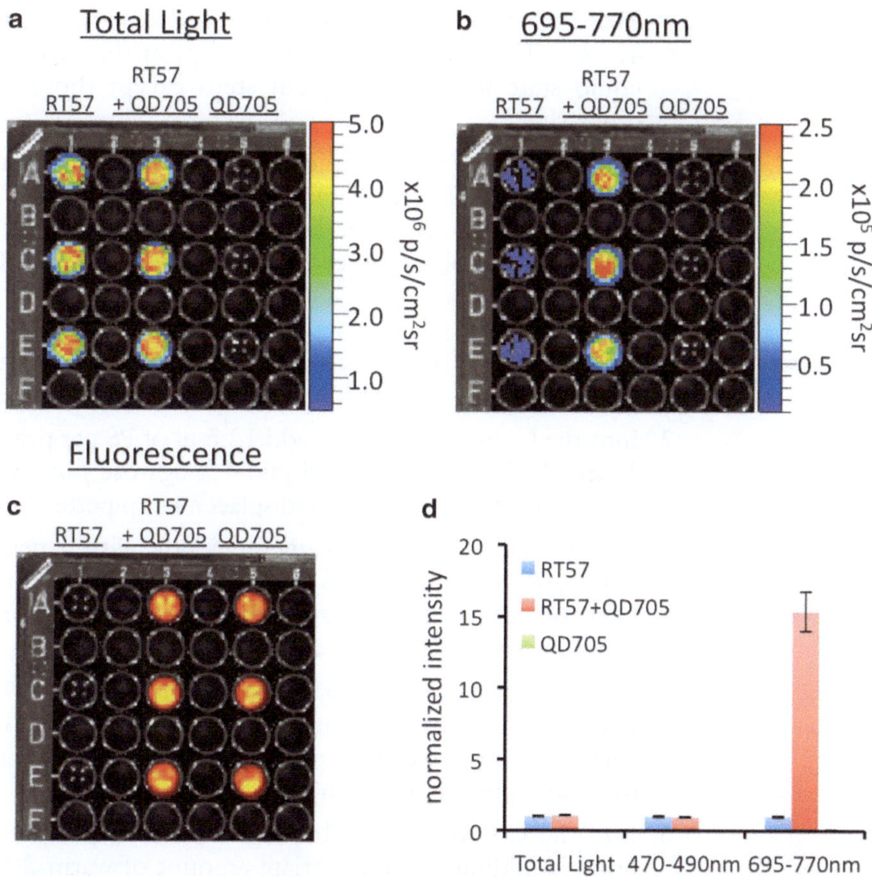

Fig. 1 A demonstration of FUEL occurring between luminescent bacteria and nanoparticles in suspension. Solutions of bacteria (Bac), bacteria with QD705 (Bac + QD705), and QD705 alone (QD705) were prepared in triplicate as described in the text and then distributed individually into a black-walled 96-well plate. The plate was then observed using an IVIS100 under the (**a**) total light, 470–490 nm band pass filter (not shown), and the (**b**) 695–770 nm band pass filter. The location of the QD705 was confirmed by epifluorescent excitation (**c**). Each individual well from (**a**) and (**b**) were normalized to the average of the Bac control under each respective filter set, with the relative intensity and standard deviation represented in (**d**), $n = 3$

3.2 FUEL Pearls Preparation

The fabrication of FUEL Pearls facilitates the transition from in vitro to in vivo conditions where the agarose provides a foundation in which the bioluminescent bacteria and the luminescent nanoparticles can remain within a constrained environment. This is advantageous the region of interest is clear and defined, and the experiment maintains conditions that are ideal for FUEL to be observed, emulating what is observed when the two components are mixed within a solution. Furthermore, the use of the agarose support allows for the creation and investigation of different FUEL configurations, providing insight into distance dependencies and overall FUEL efficiencies.

1. Prepare bacteria stock solution as previously described, however, this time aiming for an OD = 2, providing a final concentration of 1×10^9 bacteria/ml

2. Melt both the 3 % and 4 % agarose and keep it liquid (around 42 °C). The agarose needs to be kept at this temperature in liquid state for all subsequent steps except those indicating otherwise.
3. Place two 5 cm × 10 cm pieces of parafilm onto a flat surface and label one "Layered", and the other "Composite". Divide each into three equal sections using a felt pen and label one section "Control", another "Bac", and the final "Bac + QD705" (*see* **Note 4**).

3.3 Layered Pearl

1. Label tubes: "Control core"; "Control layer;" "Bac core"; "Bac layer"; "Bac + QD705 core"; "Bac + QD705 layer".
2. Into the Control core tube, add 12.5 μl of PS per pearl desired. Then add 12.5 μl of warm liquid 4 % agarose *per* pearl desired and agitate using the positive displacement pipette (*see* **Note 5**).
3. Using the positive displacement pipette with a new tip and piston, quickly dispense 25 μl aliquots of this solution onto the Control section of the parafilm. Each pearl will appear uniform and semi-spherical in shape (*see* **Note 6**).
4. Repeat the previous two steps for the remaining Bac and Bac + QD sections, replacing the 1×PS with 12.5 μl of stock bacteria (2×10^8 bacteria/ml) ensuring to use new Eppendorf tubes and pipette tips each time (*see* **Note 7**).
5. Into the Control layer tube, add 5 μl of PS per pearl desired prior to adding the appropriate amount of warm 3 % agarose (7.5 μl per pearl desired), and agitate using the positive displacement pipette. Then, using a new tip, distribute 12.5 μl aliquots onto each of previously prepared Control pearl core (*see* **Note 8**).
6. Repeat the previous step for the Bac pearls.
7. Repeat **step 5** in Subheading 3.3 for the Bac + QD pearls, this time using 5 μl of QD705 solution in place of the PS.

3.4 Composite Pearls

1. Label three Eppendorf tubes: "Control"; "Bac"; "Bac + QD705".
2. Into the Control tube add 17.5 μl of PS per pearl desired. Dispense 12.5 μl of stock bacteria (2×10^8 bacteria/ml) per pearl desired into the Bac and Bac + QD705 tubes.
3. Add 5 μl of PS or QD705 per pearl desired into the Bac or Bac + QD705 tube, respectively.
4. Using the positive displacement pipette, add 17.5 μl of warm 4 % agarose to the control tube and agitate using the pipette. Then, using a new tip, distribute 37.5 μl aliquots onto the Control section of the parafilm.
5. Repeat the previous step for the Bac and Bac + QD705, ensuring proper placement of the 37.5 μl aliquots.

3.5 Preparation and Loading of Mice

1. Before proceeding, ensure the proposed protocol adheres to the appropriate health and ethics standards for animal experimentation.
2. Using a standard 1.5 mg/kg Ketamine and 30 mg/kg Xylazine mixture, chemically anesthetize the desired number of female BALB/c mice for the experiment.
3. Using a fine electric razor carefully shave both the ventral and dorsal hind limbs of each mouse, as well as the lower abdomen, followed by cream epilation (*see* **Note 9**).
4. Place the prepared pearls into the IVIS100 and observe the bioluminescence signal under the Total Light, 470–490 nm band pass, and the 695–770 nm band pass filters. The acquisition time will be dependent upon the bioluminescent strength of the bacteria. Then acquire a fluorescence image using the 615–665 nm excitation filter and using the 695–770 nm band pass filter for the emission.
5. Make a small incision on the interior of both hind limbs of all mice (*see* **Note 10**).
6. Insert one pearl into each incision, ensuring to place one Bac pearl and one Bac+QD705 pearl per mouse. The placement of the pearls should be towards the thickest part of the thigh in order to be comparable and create the most contrast (*see* **Note 11**).
7. After insertion of the pearls, place each mouse into the IVIS100, ventral side up, separated by a piece of black paper or plastic, and commence the image acquisition (Total Light, 470–490 nm band pass, and 695–770 nm band pass). After acquisition of the luminescent signal, acquire a fluorescence image using the 615–665 nm excitation filter and a 695–770 nm emission filter.
8. Once the imaging is completed, rotate the mice to be dorsal side up, and repeat the image sequence for both the luminescence and the fluorescence. The acquisition time may need to be longer than for the ventral side (*see* Fig. 2 and **Note 12**).

3.6 In Vivo Demonstration of FUEL

To come to completion it is important to demonstrate the feasibility and the utility of the FUEL phenomenon under in vivo conditions. Building off of the previous methods, whereby FUEL was demonstrated under in vitro and constrained conditions, the final step is to remove the solid support and to observe the light output under physiologically relevant conditions. When the totality of the experiments is analyzed, a smooth and continuous relationship is found from one step to the next and a strong validation of the phenomenon is achieved. Further, the utility of FUEL is clearly shown warranting further exploration and investigation into other applications.

1. Prepare a bacterial stock solution as described in **step 1** in Subheading 3.1, resulting in a final concentration of 1×10^9 bacteria/ml.

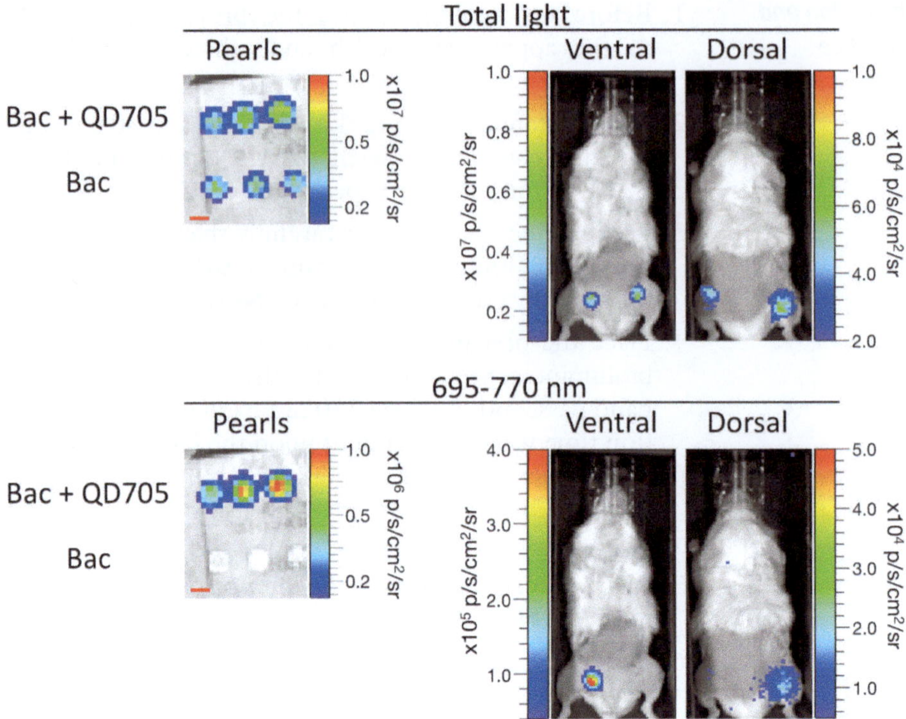

Fig. 2 Demonstration of FUEL under constrained conditions in vivo. Bac and Bac + QD705 composite Pearls were fabricated as described in the text and subsequently subcutaneously individually inserted into the inner thighs of previously prepared female BALB/c mice. The mice were placed into the IVIS100 and visualized under the Total Light (*top*), 470–490 nm band pass (*not shown*), and the 695–770 nm band pass (*bottom*) filter sets under both the ventral (*left*) and dorsal (*right*) positions. As can be seen from the figure, a substantial increase in red signal is observed from the Bac + QD705 composite Pearl compared to the Bac Pearl. This increase in red allows for the improved observation of signal through thick tissue, as is seen from the dorsal position under both filter sets

2. Label two 1.5 ml Eppendorf tubes: "Bac" and "Bac + QD705".
3. Add 20 μl of PS per mouse into the Bac tube, and 15 μl of PS per mouse into the Bac + QD705 tube (*see* **Note 13**).
4. Add 5 μl of QD705 stock per mouse into the Bac + QD705 tube.
5. As in **step 3** in Subheading 3.5, shave and epilate the hind limb and lower abdomen of each female BALB/c mouse, taking care not to injure the mice and to remove all epilation cream.
6. Add 30 μl of bacterial stock to each tube, place the tubes into the IVIS100, and launch the image sequence to verify the presence of FUEL.
7. Load one insulin syringe with the Bac solution, and a second syringe with the Bac + QD705 solution.
8. Carefully subcutaneously inject 50 μl of the Bac solution into the inner thigh of one mouse, followed quickly by subcutaneously

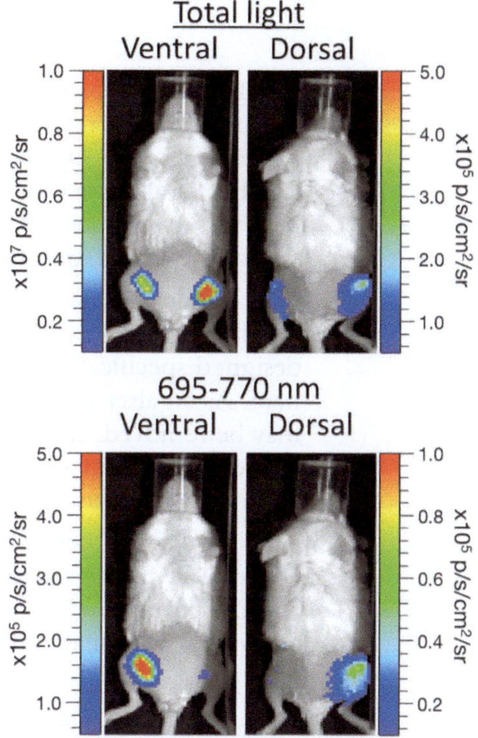

Fig. 3 In vivo demonstration of FUEL under a simulated bacterial infection. Bac and Bac + QD705 solutions were prepared as described in the text and loaded into individual insulin syringes. In a similar fashion to the Pearls, an aliquot of each solution was subcutaneously injected into the inner thighs of female BALB/c mice. The mice were then placed into the IVIS100 and visualized under the Total Light (*top*), 470–490 nm band pass (*not shown*), and the 695–770 nm band pass (*bottom*) filter sets for both the ventral (*left*) and dorsal (*right*) positions. Similar to what was observed for the Pearls, an increase of red signal was observed for the Bac + QD705 compared to the Bac alone. This increase in red signal facilitated the observation of the bacteria through thick tissue under both filter sets

injecting 50 μl of the Bac + QD705 solution into the other thigh. Rapidly repeat for the remaining mice (*see* **Note 14**).

9. After the mice are prepared and the solutions injected, place each mouse into the IVIS100, ventral side up, separated by a piece of black paper or plastic, and commence the image acquisition (Total Light, 470–490 nm band pass, and 695–770 nm band pass). After acquisition of the luminescent signal, acquire a fluorescence image using the 615–665 nm excitation filter and a 695–770 nm emission filter.

10. Once the imaging is completed, rotate the mice to be dorsal side up, and repeat the image sequence for both the luminescence and the fluorescence. The acquisition time may need to be longer than for the ventral side (*see* Fig. 3 and **Note 15**).

4 Notes

1. The RT57 were provided by José A. Bengoechea with permission from Herbert P. Schweizer [16]. Our overnight culture typically resulted in a concentration of 1.2×10^9 bacteria/ml. The bacterial strain is not overly important. We have found the best results are observed when using blue bioluminescence, similar to what is found with the *lux* operon.
2. Positive displacement pipettes greatly facilitate the fabrication of the agarose pearls. Positive displacement pipettes are designed specifically for highly viscous solutions and are ideal here. As an alternative, the distal end of a standard pipette tip may be removed, creating a larger bore to facilitate the uptake and dispensing of viscous solutions. However, this is not as accurate as the positive displacement pipettes.
3. The acquisition settings are not definitive and should be adjusted to fit the requirements of the bioluminescent source and the corresponding fluorophore that are used and the sensitivity of the available bioluminescence imaging equipment.
4. The Pearls will dry over time, so it is best to store them in a sealed container such as a Petri dish until use. Do wait more than 1 h between preparation and use.
5. Since the agarose will begin to quickly solidify, it is a good idea to prepare larger mixtures than necessary. For example, if 4 Pearls are desired, it is better to prepare for 6.
6. The agarose will coat the exterior of the pipette so be sure to insert only the distal tip of the pipette.
7. It is best to individually prepare each type of pearl, otherwise the agarose may solidify before it is distributed.
8. If this is done correctly, the layer aliquot should distribute evenly around the core, which can be verified by low magnification light microscopy. If the layer aliquot remains like a ball on top of the core, the pearl was not formed correctly. This may be due to incorrect agarose concentrations or that the agarose was too cool during distribution. For the top layer solution, the temperature can slightly elevated in order to ensure proper application.
9. While removing all the hair is not necessary, a clearer signal will be achieved due to decreased scatter. Any wounds or blisters from the razor or cream can lead to undesired background signal. Further, be sure to remove all cream from the mice, as some creams are autoluminescent leading to an increased background signal.
10. This incision should be just deep enough to allow the insertion of the pearl. Little to no blood loss should result from the incision.

11. Be sure to note if layered pearls or composite pearls are being used. Also, the placement of the pearls must be as identical as possible into each limb; otherwise the signals will not be easily comparable. The pearls must also be quite central to the limb (away from the sides) to produce quality signal. An alternative location can be the lower abdomen just above the hip joint. When viewed from the dorsal side, the tissue here is a bit thicker and quite homogeneous between the two insertion regions.
12. When placing the mice be sure to have all the mice similarly positioned. For the dorsal view, it may be necessary to position the limbs of each mouse close to its trunk to enhance the contrast.
13. Again, it is recommended to prepare for one supplementary mouse to ensure there is enough solution for the desired number of mice. For example, if three mice will be investigated, prepare the solutions for four mice.
14. The subcutaneous injection must be as compact as possible. If done correctly a small pocket of liquid should form as a result of the injection. This pocket should be stable for several minutes. Also, sometimes the hind limbs are too thin to provide good contrast. Further, 50 μl is a reliable injection volume using an insulin syringe. Clearly smaller injection volumes can be used as long as the volume is constant. An alternative location can be the lower abdomen just above the hip joint. When viewed from the dorsal side, the tissue here is a bit thicker and quite homogeneous between the two insertion regions.
15. When placing the mice be sure to have all the mice similarly positioned. For the dorsal view, it may be necessary to tuck the limbs of each mouse close to its trunk to enhance the contrast.

Acknowledgements

Joe Dragavon is a Florence Gould Scholar of the Pasteur Foundation Postdoctoral Fellowship Program. The authors would like to extend their gratitude for financial support from the Pasteur Foundation of New York (to J.D., C.S.), the EU-FP7 Program "Automation" (to S.L.S.), the Institut Carnot Program 11 (to R.T., S.L.S.) and Project IMNOS (to R.T., S.L.S.), the Conny-Maeve Charitable Foundation (S.L.S.), the European Masters in Molecular Imaging (to I.T.), the *Region Ile de France* programs MODEXA (S.L.S.), SESAME (S.L.S.), and DimMalInf (S.L.S., R.T.), and the Institut Pasteur, Paris. Further, the authors would like to thank Bruno Baron of the *Plate-Forme de Biophysique des Macromolécules et de leurs Interactions* and Marie-Anne Nicola of the *Plate-Forme d'Imagerie Dynamique* for technical support and assistance, José Bengoechea and Herbert Schweizer for reagents, and Philippe Sansonetti for use of lab space, reagents, and equipment.

References

1. Contag CH, Ross BD (2002) It's not just about anatomy: in vivo bioluminescence imaging as an eyepiece into biology. J Magn Reson Imaging 16:378–387
2. Dragavon J, Blazquez S, Rekiki A et al (2012) In vivo excitation of nanoparticles using luminescent bacteria. Proc Natl Acad Sci USA 109: 8890–8895
3. Choy G, O'Connor S, Diehn FE et al (2003) Comparison of noninvasive fluorescent and bioluminescent small animal optical imaging. Biotechniques 35:1022–1030
4. Badr CE, Tannous BA (2011) Bioluminescence imaging: progress and applications. Trends Biotechnol 29:624–633
5. Zinn KR, Chaudhuri TR, Szafran AA et al (2008) Noninvasive bioluminescence imaging in small animals. ILAR J 49:103–115
6. Madero-Visbal RA, Colon JF, Hernandez IC et al (2010) Bioluminescence imaging correlates with tumor progression in an orthotopic mouse model of lung cancer. Surg Oncol 21: 23–29
7. Warawa JM, Long D, Rosenke R et al (2011) Bioluminescent diagnostic imaging to characterize altered respiratory tract colonization by the burkholderia pseudomallei capsule mutant. Front Microbiol 2:133
8. Roda A, Guardigli M (2011) Analytical chemiluminescence and bioluminescence: latest achievements and new horizons. Anal Bioanal Chem 402:69–76
9. Kielland A, Blom T, Nandakumar KS et al (2009) In vivo imaging of reactive oxygen and nitrogen species in inflammation using the luminescent probe L-012. Free Radic Biol Med 47:760–766
10. Dragavon J, Blazquez S, Rogers K et al (2011) Validation of method for enhanced production of red-shifted bioluminescent photons in vivo. Imaging, manipulation, and analysis of biomolecules, cells, and tissues IX, Vol. 7902. (eds. D.L. Farkas, D.V. Nicolau & R.C. Leif) 790210-790219 (International society for optics and photonics, 2011).
11. Colin M, Moritz S, Schneider H et al (2000) Haemoglobin interferes with the ex vivo luciferase luminescence assay: consequence for detection of luciferase reporter gene expression in vivo. Gene Ther 7:1333–1336
12. Branchini BR, Ablamsky DM, Rosenberg JC (2010) Chemically modified firefly luciferase is an efficient source of near-infrared light. Bioconjug Chem 21:2023–2030
13. Branchini BR, Rosenberg JC, Ablamsky DM et al (2011) Sequential bioluminescence resonance energy transfer-fluorescence resonance energy transfer-based ratiometric protease assays with fusion proteins of firefly luciferase and red fluorescent protein. Anal Biochem 414: 239–245
14. So M-K, Xu C, Loening AM et al (2006) Self-illuminating quantum dot conjugates for in vivo imaging. Nat Biotechnol 24: 339–343
15. Xiong L, Shuhendler AJ, Rao J (2012) Self-luminescing BRET-FRET near-infrared dots for in vivo lymph-node mapping and tumour imaging. Nat Commun 3:1193
16. Choi K-H, Schweizer HP (2006) mini-Tn7 insertion in bacteria with single attTn7 sites: example Pseudomonas aeruginosa. Nat Protoc 1:153–161

Index

Christian E. Badr (ed.), *Bioluminescent Imaging: Methods and Protocols*, Methods in Molecular Biology, vol. 1098,
DOI 10.1007/978-1-62703-718-1, © Springer Science+Business Media New York 2014

C

D

E

MIX
Papier aus verantwortungsvollen Quellen
Paper from responsible sources
FSC® C105338

If you have any concerns about our products, you can contact us on
ProductSafety@springernature.com

In case Publisher is established outside the EU, the EU authorized representative is:
Springer Nature Customer Service Center GmbH
Europaplatz 3, 69115 Heidelberg, Germany

Printed by Libri Plureos GmbH
in Hamburg, Germany